影 印 版 说 明

MOMENTUM PRESS 出版的 *Plastics Technology Handbook*（2 卷）是介绍塑料知识与技术的大型综合性手册，内容涵盖了从高分子基本原理，到塑料的合成、种类、性能、配料、加工、制品，以及模具、二次加工等各个方面。通过阅读、学习本手册，无论是专业人员还是非专业人员，都会很快熟悉和掌握塑料制品的设计和制造方法。可以说一册在手，别无他求。

原版 2 卷影印时分为 11 册，第 1 卷分为：
  塑料基础知识·塑料性能
  塑料制品生产
  注射成型
  挤压成型
  吹塑成型
  热成型
  发泡成型·压延成型
第 2 卷分为：
  涂层·浇注成型·反应注射成型·旋转成型
  压缩成型·增强塑料·其他工艺
  模具
  辅机与二次加工设备

  唐纳德 V·罗萨多，波士顿大学化学学士学位，美国东北大学 MBA 学位，马萨诸塞大学洛厄尔分校工程塑料和加州大学工商管理博士学位（伯克利）。著有诸多论文及著作，包括《塑料简明百科全书》、《注塑手册（第三版）》以及塑料产品材料和工艺选择手册等。活跃于塑料界几十年，现任著名的 Plasti Source Inc. 公司总裁，并是美国塑料工业协会（SPI）、美国塑料学会（PIA）和 SAMPE（The Society for the Advancement of Material and Process Engineering）的重要成员。

材料科学与工程图书工作室
  联系电话 0451-86412421
      0451-86414559
  邮 箱 yh_bj@aliyun.com
      xuyaying81823@gmail.com
      zhxh6414559@aliyun.com

影印版

PLASTICS TECHNOLOGY HANDBOOK

塑料技术手册

VOLUME 2

AUXILIARY AND SECONDARY EQUIPMENT

辅机与二次加工设备

EDITED BY

DONALD V. ROSATO
MARLENE G. ROSATO
NICK R. SCHOTT

哈爾濱工業大學出版社
HARBIN INSTITUTE OF TECHNOLOGY PRESS

黑版贸审字08-2014-093号

Donald V.Rosato, Marlene G.Rosato, Nick R.Schott
Plastics Technology Handbook Volume 2
9781606500828
Copyright © 2012 by Momentum Press, LLC
All rights reserved.

Originally published by Momentum Press, LLC
English reprint rights arranged with Momentum Press, LLC through McGraw-Hill Education (Asia)

This edition is authorized for sale in the People's Republic of China only, excluding Hong Kong, Macao SAR and Taiwan.

本书封面贴有McGraw-Hill Education公司防伪标签，无标签者不得销售。
版权所有，侵权必究。

图书在版编目（CIP）数据

塑料技术手册. 第2卷. 辅机与二次加工设备 =Plastics technology handbook volume 2 auxiliary and secondary equipment : 英文 /（美）罗萨多（Rosato, D. V.）等主编. —影印本. — 哈尔滨 : 哈尔滨工业大学出版社, 2015.6
　　ISBN 978-7-5603-5052-3

　　Ⅰ. ①塑… Ⅱ. ①罗… Ⅲ. ①塑料－技术手册－英文 ②塑料成型加工设备－技术手册－英文 Ⅳ. ①TQ320.6-62

中国版本图书馆CIP数据核字（2014）第280099号

责任编辑	许雅莹　张秀华　杨　桦
出版发行	哈尔滨工业大学出版社
社　　址	哈尔滨市南岗区复华四道街10号　邮编150006
传　　真	0451-86414749
网　　址	http://hitpress.hit.edu.cn
印　　刷	哈尔滨市石桥印务有限公司
开　　本	787mm×960mm　1/16　印张16.25
版　　次	2015年6月第1版　2015年6月第1次印刷
书　　号	ISBN 978-7-5603-5052-3
定　　价	84.00元

（如因印刷质量问题影响阅读，我社负责调换）

PLASTICS TECHNOLOGY HANDBOOK

VOLUME 2

EDITED BY

Donald V. Rosato, PhD, MBA, MS, BS, PE
PlastiSource Inc.
Society of Plastics Engineers
Plastics Pioneers Association
UMASS Lowell Plastics Advisory Board

Marlene G. Rosato, BASc (ChE), P Eng
Gander International Inc.
Canadian Society of Chemical Engineers
Association of Professional Engineers of Ontario
Product Development and Management Association

Nick R. Schott, PhD, MS, BS (ChE), PE
UMASS Lowell Professor of Plastics Engineering Emeritus & Plastics Department Head Retired
Plastics Institute of America
Secretary & Director for Educational and Research Programs

Momentum Press, LLC, New York

Contents

FIGURES 9

TABLES 13

ABBREVIATIONS 16

ACKNOWLEDGMENTS 24

PREFACE 25

ABOUT THE AUTHORS 28

18. AUXILIARY AND SECONDARY EQUIPMENT 738
    INTRODUCTION 738
    MATERIAL/PRODUCT HANDLING 756
        Material-Handling System 757
        Injection Molding 777
        Extruding 786
    DECORATING 805
    JOINING AND ASSEMBLING 807
        Adhesive and Solvent Bonding 807
        Mechanical Assembly 835
        Staking 849
        Welding Assembly 863
    MACHINING 892

Overview	892
Machining and Cutting Operations	897
Machining and Tooling	911
Machining Nonmelt TP	919
Laser Machining	922
Other Machining Methods	923
Machining Safety	924

GLOSSARY 925

FURTHER READING 953

Figures

Figure 18.1	Example of AE required for plastics going from a railcar to a silo.	744
Figure 18.2	Closeup view of a piping system to and from silos, with each having a capacity of 2000 lb.	745
Figure 18.3	Examples of plant layout with extrusion and injection molding primary and AE.	746
Figure 18.4	Example of an extrusion laminator with AE.	747
Figure 18.5	Example of a blow-molding extruder with AE (rolls, turret winder, etc.).	748
Figure 18.6	Example of an extruder coater with AE.	749
Figure 18.7	Example of plant layout with injection molding primary and AE.	749
Figure 18.8	Example of extruded products requiring AE.	750
Figure 18.9	Example of ventilation AE used with an injection molding machine (courtesy of Husky Injection Molding Systems Inc.).	751
Figure 18.10	Examples of material handling AE used with an injection molding machine (courtesy of Husky Injection Molding Systems Inc.).	752
Figure 18.11	Example of a pneumatic vacuum venturi flow system.	757
Figure 18.12	Example of continuous pressure pellets with rates based on polystyrene at 35 lb/ft^3 (560 kg/m^3).	760
Figure 18.13	Example of continuous vacuum pellets with rates based on polystyrene at 35 lb/ft^3 (560 kg/m^3).	761
Figure 18.14	Example of continuous vacuum powder with rates based on polyvinyl chloride (PVC) at 35 lb/ft^3 (560 kg/m^3).	762
Figure 18.15	Example of a 10 hp vacuum system conveying polystyrene at 35 lb/ft^3 (560 kg/m^3).	763

Figure 18.16	Example of a 25 hp vacuum system conveying polystyrene at 35 lb/ft^3 (560 kg/m^3).	764
Figure 18.17	Example of a single pneumatic material-handling line-feeding hoppers.	768
Figure 18.18	Example of the front and side views of a basic hopper.	769
Figure 18.19	Introduction to hopper mixers.	770
Figure 18.20	Example of a dump-type hopper loader.	770
Figure 18.21	Example of a screw-controlled feeding loader (courtesy of Spirex Corporation).	771
Figure 18.22	Detail view of a hopper screw-controlled feeding loader.	771
Figure 18.23	Example of components in a hopper blender.	772
Figure 18.24	Example of metering a color additive in a blender.	773
Figure 18.25	Example of a hopper power-pump loader.	773
Figure 18.26	Example of a vacuum hopper-loading cycle.	774
Figure 18.27	Systems utilizing a rotary air lock feeder to separate pressure and vacuum airflow.	775
Figure 18.28	Examples of coarse, dusty, and powder material-filtering systems.	776
Figure 18.29	Example of a positive take-out and transfer mechanism for molded products (courtesy of Husky Injection Molding Systems Inc.).	778
Figure 18.30	Example of a positive take-out system to handle and pack molded products (courtesy of Husky Injection Molding Systems Inc.).	779
Figure 18.31	Example of a free-drop take-out and transfer mechanism of molded products.	780
Figure 18.32	Example of an unscramble-and-orient system for molded products (courtesy of Husky Injection Molding Systems Inc.).	781
Figure 18.33	Example of bulk filling with automatic carton indexing of molded products (courtesy of Husky Injection Molding Systems Inc.).	781
Figure 18.34	Example of flow of material to shipping of molded products.	782
Figure 18.35	Example of a robot removing parts from a mold and depositing them in orderly fashion in a container.	783
Figure 18.36	Mold base en route manually to injection molding press.	788
Figure 18.37	Mold base placed manually to the right in injection molding press.	789
Figure 18.38	Fully automatic horizontal mold change (courtesy of Staubli Corp., Duncan, South Carolina).	790
Figure 18.39	Fully automatic overhead-crane mold change.	790
Figure 18.40	Examples of tension-control rollers in a film, sheet, or coating line.	791
Figure 18.41	Example of laminating with an adhesive.	791
Figure 18.42	Example of roll-change-sequence winder (courtesy of Black Clawson).	791
Figure 18.43	Closeup view of a tension roll that is processing plastic film.	792
Figure 18.44	Example herringbone idler reducing wrinkles of web.	792
Figure 18.45	Examples of drum-cooling designs with shell cooling being the best design.	793

Figure 18.46	Examples of matted and unmatted embossing rolls.	793
Figure 18.47	Example of a wood-grain embossing roll.	794
Figure 18.48	Example of ultrasonically sealing a decorative pattern.	794
Figure 18.50	Example of a dancer roll controlling tension in an extruded sheet line.	795
Figure 18.51	Example of an extruded sheet line turret wind-up reel change system.	795
Figure 18.49	Guide to sheet-polishing roll sizes with a 450°F (230°C) melt temperature.	795
Figure 18.52	View of a large single winder at the end of an extruder sheet line (courtesy of Welex).	796
Figure 18.53	View of a large dual-turret winder at the end of an extruder sheet line.	797
Figure 18.54	View of a sheet roll stock extruder winder with triple fixed shafts (courtesy of Welex).	798
Figure 18.55	View of downstream extruder-blown film line going through control rolls and dual wind-up turrets (courtesy of Windmoeller & Hoelscher Corporation).	799
Figure 18.56	Examples of pipe-extrusion caterpillar puller with rollers and conveyor belts.	800
Figure 18.57	Description of a caterpillar belt puller used in an extruder line (courtesy of Conair).	801
Figure 18.58	Description of a vacuum sizing tank used in an extruder line (courtesy of Conair).	801
Figure 18.59	Description of a water-and-spray tank used in an extruder line (courtesy of Conair).	802
Figure 18.60	Description of a rotary knife cutter used in an extruder line (courtesy of Conair).	802
Figure 18.61	Description of a pneumatic-stop rotary knife cutter used in an extruder line (courtesy of Conair).	803
Figure 18.62	Description of a traveling up-cut saw used in an extruder line (courtesy of Conair).	803
Figure 18.63	Description of a product takeaway conveyor used in an extruder line (courtesy of Conair)	804
Figure 18.64	Examples in the use of masking for paint spraying.	814
Figure 18.65	Examples of paint spray-and-wipe.	815
Figure 18.66	Examples of screen printing.	815
Figure 18.67	Example of hot stamping using a roll-on technique.	815
Figure 18.68	Example of pad transfer printing.	816
Figure 18.69	Joining and bonding methods.	830
Figure 18.70	Examples of joint geometries.	831
Figure 18.71	Examples of corona treatments in extrusion lines.	839
Figure 18.72	Guide for molding threads.	852
Figure 18.73	Examples of assembling all plastic and plastic to different materials where thermal stresses can become a problem when proper design is not used (chapter 19).	853

Figure 18.74	Examples of self-tapping screws.	855
Figure 18.75	Molded-in insert designs.	856
Figure 18.76	Examples of metal-expansion types of slotted and nonslotted inserts.	859
Figure 18.77	Examples of press-fit-stress analyses (courtesy of Bayer).	861
Figure 18.78	Examples of cantilever beam snap-fits.	863
Figure 18.79	Example of cold staking of plastic.	864
Figure 18.80	Example of hot staking of plastic.	864
Figure 18.81	Example of hot-plate welding.	869
Figure 18.82	Film-welded, 8-ply arrangement using a Doboy thermal welder.	872
Figure 18.83	Example of a manual hot-gas welding.	874
Figure 18.84	Example of an automatic hot-gas welder; hot gas blown between sheets, which melt and flow together.	874
Figure 18.85	Example of design joints for hot-gas welding.	875
Figure 18.86	Examples of visually examining hot-gas weld quality.	875
Figure 18.87	Example of linear-vibration welding.	876
Figure 18.88	Penetration-versus-time curve showing the four phases of vibration welding.	876
Figure 18.89	Spin welding, where one part does not move and the other part rotates.	881
Figure 18.90	Example of a joint used in spin welding.	881
Figure 18.91	Components of an ultrasonic welder.	882
Figure 18.92	Stages in ultrasonic welding.	883
Figure 18.93	Examples of plastic mating joints to be ultrasonically welded.	884
Figure 18.94	Example of induction heat produced during induction welding.	886
Figure 18.95	Example of induction welding a lid to a container.	886
Figure 18.96	The three steps in resistance welding.	890
Figure 18.97	Example of an extrusion-welding system, where the hot air melts the plastic to be welded prior to the extruded melt flows into the area.	891
Figure 18.98	Examples of cutting and punching in-line, extruded TPs.	895
Figure 18.99	Example of extrusion in-line shear cutter with sheets being stacked.	897
Figure 18.100	Guide to slitting extruded film or coating.	909
Figure 18.101	Schematics of cutting-tool actions.	911
Figure 18.102	Basic schematic of a cutting tool.	913
Figure 18.103	Example of forces acting on a tool.	914
Figure 18.104	Example of wear pattern.	915
Figure 18.105	Nomenclature for single-point tools.	918
Figure 18.106	Nomenclature of twist drills.	918
Figure 18.107	Nomenclature of milling cutters.	919
Figure 18.108	Cutting tool for machining (skiving) tape from a molded plastic block.	922

Tables

Table 18.1	Example of manufacturing cycle that includes equipment	739
Table 18.2	SPE auxiliaries buyer's guide (courtesy of SPE)	740
Table 18.3	Introduction to auxiliary and SE performances	754
Table 18.4	Examples of auxiliary and SE	755
Table 18.5	Estimated annual savings for energy-efficient electric motors (Electrical Apparatus Service Association)	765
Table 18.6	Examples of the usual functions of robots and perimeter guarding	784
Table 18.7	Examples of comparing robots with other parts-handling systems	786
Table 18.8	Examples of types of robots manufactured	787
Table 18.9	Examples of different rolls used in different extrusion processes	806
Table 18.10	Guide to decorating	808
Table 18.11	Examples of methods for decorating plastic products after fabrication	810
Table 18.12	Examples of methods for decorating plastic products in a mold	811
Table 18.13	Guide in comparing a few decorating methods from size to cost	812
Table 18.14	Review of a few decorating methods	813
Table 18.15	Examples of joining methods	817
Table 18.16	Examples of joining TPs and TSs	817
Table 18.17	Examples of descriptions for different joining methods	818
Table 18.18	Directory of companies that provide joining and assembling methods	820
Table 18.19	Examples of adhesives for bonding plastics to plastics	826
Table 18.20	Examples of bonding TPs to nonplastics	829
Table 18.21	Examples of bonding TS plastics to nonplastics	829
Table 18.22	Adhesive terminology	832
Table 18.23	Example of adhesives classified by composition	834
Table 18.24	Plasma treatment	836
Table 18.26	Peel strength of plastics after plasma treatment per ASTM test methods	837
Table 18.25	Lap shear strength of plastics after plasma treatment per American Society for Testing Materials (ASTM) test methods	837

Table 18.27	Shear strength of PP to PP adhesive bonds in psi (MPa) per ASTM D 4501	838
Table 18.28	Shear strength of polyethylene (PE) to PE in psi (MPa)	840
Table 18.29	Shear strength of ABS to ABS in psi (MPa)	841
Table 18.30	Shear strength of PP to PP in psi (MPa)	842
Table 18.31	Shear strength of PVC to PVC in psi (MPa)	843
Table 18.32	Shear strength of polycarbonate (PC) to PC in psi (MPa)	844
Table 18.33	Shear strength of PUR to PUR in psi (MPa)	845
Table 18.34	Shear strength of PA to PA in psi (MPa)	846
Table 18.35	Shear strength of polyimide to polyimide in psi (MPa)	847
Table 18.36	Shear strength of acetal to acetal in psi (MPa)	848
Table 18.37	Shear strength of polymethyl methacrylate (PMMA) to PMMA in psi (MPa)	849
Table 18.38	Shear strength of polyethylene terephthalate (PET) to PET in psi (MPa)	850
Table 18.39	Shear strength of polyetheretherketone (PEEK) to PEEK in psi (MPa)	850
Table 18.40	Shear strength of liquid crystal polymer (LCP) to LCP in psi (MPa)	851
Table 18.41	Shear strength of fluoroplastic to fluoroplastic in psi (MPa)	851
Table 18.42	Guide relating molded wall thicknesses to insert diameters (in [mm])	862
Table 18.43	Examples of welding methods versus tensile-strength retention	865
Table 18.44	Examples of welding characteristics	865
Table 18.45	Examples of ultrasonic welding applications	866
Table 18.46	Comparison of a few welding methods	866
Table 18.47	Comparing welding of different plastics, each to itself	867
Table 18.48	Economic guide to a few welding processes	868
Table 18.49	Tensile strength of hot-plate welding PP copolymerized with ethylene pipe	870
Table 18.50	Impact and tensile strength of hot-plate welding high-density polyethylene (HDPE)	870
Table 18.51	Tensile strength of different hot-plate welds of PP copolymerized with ethylene pipe	870
Table 18.52	Tensile strength of hot-plate welding ABS	871
Table 18.53	Properties of vibration welds of PC to itself and other plastics	877
Table 18.54	Properties of vibration welds of PC/ABS to itself and other plastics	877
Table 18.55	Properties of vibration welds of PC/polybutylene terephthalate (PBT) to itself and to PC	878
Table 18.56	Properties of vibration welds of ABS to itself and other plastics	878
Table 18.57	Properties of vibration welds of acrylonitrile-styrene-acrylate (ASA) to itself	879
Table 18.58	Properties of vibration welds of PS-modified PPE/PA to itself and other plastics	879
Table 18.59	Properties of vibration welds of modified polypropylene oxide (PPO) to itself and other plastics	880

Table 18.60	Properties of vibration welds of PBT to itself and other plastics	880
Table 18.61	Example of a boss-hole design for the use of ultrasonically installed inserts using styrene maleic anhydride copolymer	884
Table 18.62	Optimum ultrasonic welding conditions for impact-modified PET-PC blend	884
Table 18.63	Weld strength of ultrasonic bonds of medical plastics; three letters in each box represent bonds subjected to no sterilization, ethylene-oxide sterilization, and gamma-radiation sterilization, respectively	885
Table 18.64	Guide to bonding plastic to plastic via induction welding	886
Table 18.65	Properties of radio-frequency welding of flexible PVC to itself and other plastics	888
Table 18.66	Properties of radio-frequency welding of rigid PVC to itself and other plastics	889
Table 18.67	Properties of radio-frequency welding of aromatic polyester PUR to itself and other plastics	889
Table 18.68	Properties of laser-welded PE joints	892
Table 18.69	Properties of laser-welded PP joints	892
Table 18.70	Examples of machining operations	893
Table 18.71	Examples of finishing operations	893
Table 18.72	Examples of supplementary machining operations	894
Table 18.73	Guide to single-point box-tool machining (chapter 17 reviews tool materials)	898
Table 18.74	Guide to turning, cutoff, and form-tool machining	899
Table 18.75	Guide to drilling	900
Table 18.76	Guide to end milling: Slotting machining	901
Table 18.77	Guide to end milling: Peripheral machining	902
Table 18.78	Guide to side and slot milling arbor-mounted cutter machining	903
Table 18.79	Guide to face-milling machining	904
Table 18.80	Guide to power band sawing	905
Table 18.81	Guide to tapping TPs and TS plastics	905
Table 18.82	Guide to reaming TPs and TS plastics	906
Table 18.83	Guide to standard tolerances for punched holes and slots in sheet stock	907
Table 18.84	NEMA guide to standard tolerances for punched holes and slots in high-pressure composite laminated grades of sheet stock, rods, and tubes	908
Table 18.85	Guide to cutting equipment capabilities	908
Table 18.86	Guide to drill geometry	908
Table 18.87	Examples of cutting-tool geometries	912
Table 18.88	Guide for drilling ½ to ⅜ in holes in TPs	919

Abbreviations

AA acrylic acid
AAE American Association of Engineers
AAES American Association of Engineering Societies
ABR polyacrylate
ABS acrylontrile-butadiene-styrene
AC alternating current
ACS American Chemical Society
ACTC Advanced Composite Technology Consortium
ad adhesive
ADC allyl diglycol carbonate (also CR-39)
AFCMA Aluminum Foil Container Manufacturers' Association
AFMA American Furniture Manufacturers' Association
AFML Air Force Material Laboratory
AFPA American Forest and Paper Association
AFPR Association of Foam Packaging Recyclers
AGMA American Gear Manufacturers' Association
AIAA American Institute of Aeronautics and Astronauts
AIChE American Institute of Chemical Engineers
AIMCAL Association of Industrial Metallizers, Coaters, and Laminators
AISI American Iron and Steel Institute
AMBA American Mold Builders Association
AMC alkyd molding compound
AN acrylonitrile
ANSI American National Standards Institute
ANTEC Annual Technical Conference (of the Society of the Plastic Engineers)
APC American Plastics Council
APET amorphous polyethylene terephthalate
APF Association of Plastics Fabricators
API American Paper Institute
APME Association of Plastics Manufacturers in Europe
APPR Association of Post-Consumer Plastics Recyclers
AQL acceptable quality level
AR aramid fiber; aspect ratio
ARP advanced reinforced plastic
ASA acrylonitrile-styrene-acrylate
ASCII american standard code for information exchange
ASM American Society for Metals

ASME American Society of Mechanical Engineers
ASNDT American Society for Non-Destructive Testing
ASQC American Society for Quality Control
ASTM American Society for Testing Materials
atm atmosphere
bbl barrel
BFRL Building and Fire Research Laboratory
Bhn Brinell hardness number
BM blow molding
BMC bulk molding compound
BO biaxially oriented
BOPP biaxially oriented polypropylene
BR polybutadiene
Btu British thermal unit
buna polybutadiene
butyl butyl rubber
CA cellulose acetate
CAB cellulose acetate butyrate
CaCO$_3$ calcium carbonate (lime)
CAD computer-aided design
CAE computer-aided engineering
CAM computer-aided manufacturing
CAMPUS computer-aided material preselection by uniform standards
CAN cellulose acetate nitrate
CAP cellulose acetate propionate
CAS Chemical Abstract Service (a division of the American Chemical Society)
CAT computer-aided testing
CBA chemical blowing agent
CCA cellular cellulose acetate
CCV Chrysler composites vehicle
CEM Consorzio Export Mouldex (Italian)
CFA Composites Fabricators Association
CFC chlorofluorocarbon
CFE polychlorotrifluoroethylene
CIM ceramic injection molding; computer integrated manufacturing
CLTE coefficient of linear thermal expansion
CM compression molding
CMA Chemical Manufacturers' Association
CMRA Chemical Marketing Research Association
CN cellulose nitrate (celluloid)
CNC computer numerically controlled
CP Canadian Plastics
CPE chlorinated polyethylene
CPET crystallized polyethylene terephthalate
CPI Canadian Plastics Institute
cpm cycles/minute
CPVC chlorinated polyvinyl chloride
CR chloroprene rubber; compression ratio
CR-39 allyl diglycol carbonate
CRP carbon reinforced plastics
CRT cathode ray tube
CSM chlorosulfonyl polyethylene
CTFE chlorotrifluorethylene
DAP diallyl phthalate
dB decibel
DC direct current
DEHP diethylhexyl phthalate
den denier
DGA differential gravimetric analysis
DINP diisononyl phthalate
DMA dynamic mechanical analysis
DMC dough molding compound
DN *Design News* publication
DOE Design of Experments
DSC differential scanning calorimeter
DSD Duales System Deutschland (German Recycling System)
DSQ German Society for Quality
DTA differential thermal analysis
DTGA differential thermogravimetric analysis
DTMA dynamic thermomechanical analysis
DTUL deflection temperature under load
DV devolatilization
DVR design value resource; dimensional velocity research; Druckverformungsrest (German

compression set); dynamic value research; dynamic velocity ratio
E modulus of elasticity; Young's modulus
EBM extrusion blow molding
E_c modulus, creep (apparent)
EC ethyl cellulose
ECTFE polyethylene-chlorotrifluoroethylene
EDM electrical discharge machining
E/E electronic/electrical
EEC European Economic Community
EI modulus × moment of inertia (equals stiffness)
EMI electromagnetic interference
EO ethylene oxide (also EtO)
EOT ethylene ether polysulfide
EP ethylene-propylene
EPA Environmental Protection Agency
EPDM ethylene-propylene diene monomer
EPM ethylene-propylene fluorinated
EPP expandable polypropylene
EPR ethylene-propylene rubber
EPS expandable polystyrene
E_r modulus, relaxation
E_s modulus, secant
ESC environmental stress cracking
ESCR environmental stress cracking resistance
ESD electrostatic safe discharge
ET ethylene polysulfide
ETFE ethylene terafluoroethylene
ETO ethylene oxide
EU entropy unit; European Union
EUPC European Association of Plastics Converters
EUPE European Union of Packaging and Environment
EUROMAP Eu^ropean Committee of Machine Manufacturers for the Rubber and Plastics Industries (Zurich, Switzerland)
EVA ethylene-vinyl acetate
E/VAC ethylene/vinyl acetate copolymer
EVAL ethylene-vinyl alcohol copolymer (tradename for EVOH)
EVE ethylene-vinyl ether
EVOH ethylene-vinyl alcohol copolymer (or EVAL)
EX extrusion
F coefficient of friction; Farad; force
FALLO follow all opportunities
FDA Food and Drug Administration
FEA finite element analysis
FEP fluorinated ethylene-propylene
FFS form, fill, and seal
FLC fuzzy logic control
FMCT fusible metal core technology
FPC flexible printed circuit
fpm feet per minute
FRCA Fire Retardant Chemicals Association
FRP fiber reinforced plastic
FRTP fiber reinforced thermoplastic
FRTS fiber reinforced thermoset
FS fluorosilicone
FTIR Fourier transformation infrared
FV frictional force × velocity
G gravity; shear modulus (modulus of rigidity); torsional modulus
GAIM gas-assisted injection molding
gal gallon
GB gigabyte (billion bytes)
GD&T geometric dimensioning and tolerancing
GDP gross domestic product
GFRP glass fiber reinforced plastic
GMP good manufacturing practice
GNP gross national product
GP general purpose
GPa giga-Pascal
GPC gel permeation chromatography
gpd grams per denier
gpm gallons per minute
GPPS general purpose polystyrene
GRP glass reinforced plastic
GR-S polybutadiene-styrene
GSC gas solid chromatography

H hysteresis; hydrogen
HA hydroxyapatite
HAF high-abrasion furnace
HB Brinell hardness number
HCFC hydrochlorofluorocarbon
HCl hydrogen chloride
HDPE high-density polyethylene (also PE-HD)
HDT heat deflection temperature
HIPS high-impact polystyrene
HMC high-strength molding compound
HMW-HDPE high molecular weight–high density polyethylene
H-P Hagen-Poiseuille
HPLC high-pressure liquid chromatography
HPM hot pressure molding
HTS high-temperature superconductor
Hz Hertz (cycles)
I integral; moment of inertia
IB isobutylene
IBC internal bubble cooling
IBM injection blow molding; International Business Machines
IC *Industrial Computing* publication
ICM injection-compression molding
ID internal diameter
IEC International Electrochemical Commission
IEEE Institute of Electrical and Electronics Engineers
IGA isothermal gravimetric analysis
IGC inverse gas chromatography
IIE Institute of Industrial Engineers
IM injection molding
IMM injection molding machine
IMPS impact polystyrene
I/O input/output
ipm inch per minute
ips inch per second
IR synthetic polyisoprene (synthetic natural rubber)
ISA Instrumentation, Systems, and Automation
ISO International Standardization Organization or International Organization for Standardization
IT information technology
IUPAC International Union of Pure and Applied Chemistry
IV intrinsic viscosity
IVD in vitro diagnostic
J joule
JIS Japanese Industrial Standard
JIT just-in-time
JIT just-in-tolerance
J$_p$ polar moment of inertia
JSR Japanese SBR
JSW Japan Steel Works
JUSE Japanese Union of Science and Engineering
JWTE Japan Weathering Test Center
K bulk modulus of elasticity; coefficient of thermal conductivity; Kelvin; Kunststoffe (plastic in German)
kb kilobyte (1000 bytes)
kc kilocycle
kg kilogram
KISS keep it short and simple
Km kilometer
kPa kilo-Pascal
ksi thousand pounds per square inch (psi $\times$ 10^3)
lbf pound-force
LC liquid chromatography
LCP liquid crystal polymer
L/D length-to-diameter (ratio)
LDPE low-density polyethylene (PE-LD)
LIM liquid impingement molding; liquid injection molding
LLDPE linear low-density polyethylene (also PE-LLD)
LMDPE linear medium density polyethylene
LOX liquid oxygen
LPM low-pressure molding
m matrix; metallocene (catalyst); meter

mμ micromillimeter; millicron; 0.000001 mm
μm micrometer
MA maleic anhydride
MAD mean absolute deviation; molding area diagram
Mb bending moment
MBTS benzothiazyl disulfide
MD machine direction; mean deviation
MD&DI Medical Device and Diagnostic Industry
MDI methane diisocyanate
MDPE medium density polyethylene
Me metallocene catalyst
MF melamine formaldehyde
MFI melt flow index
mHDPE metallocene high-density polyethylene
MI melt index
MIM metal powder injection molding
MIPS medium impact polystyrene
MIT Massachusetts Institute of Technology
mLLDPE metallocene catalyst linear low-density polyethylene
MMP multimaterial molding or multimaterial multiprocess
MPa mega-Pascal
MRPMA Malaysian Rubber Products Manufacturers' Association
Msi million pounds per square inch (psi $\times 10^6$)
MSW municipal solid waste
MVD molding volume diagram
MVT moisture vapor transmission
MW molecular weight
MWD molecular weight distribution
MWR molding with rotation
N Newton (force)
NACE National Association of Corrosion Engineers
NACO National Association of CAD/CAM Operation
NAGS North America Geosynthetics Society
NASA National Aeronautics Space Administration
NBR butadiene acrylontrile
NBS National Bureau of Standards (since 1980 renamed the National Institute Standards and Technology or NIST)
NC numerical control
NCP National Certification in Plastics
NDE nondestructive evaluation
NDI nondestructive inspection
NDT nondestructive testing
NEAT nothing else added to it
NEMA National Electrical Manufacturers' Association
NEN Dutch standard
NFPA National Fire Protection Association
NISO National Information Standards Organization
NIST National Institute of Standards and Technology
nm nanometer
NOS not otherwise specified
NPCM National Plastics Center and Museum
NPE National Plastics Exhibition
NPFC National Publications and Forms Center (US government)
NR natural rubber (polyisoprene)
NSC National Safety Council
NTMA National Tool and Machining Association
NWPCA National Wooden Pallet and Container Association
OD outside diameter
OEM original equipment manufacturer
OPET oriented polyethylene terephthalate
OPS oriented polystyrene
OSHA Occupational Safety and Health Administration
P load; poise; pressure
Pa Pascal
PA polyamide (nylon)
PAI polyamide-imide
PAN polyacrylonitrile

PB polybutylene
PBA physical blowing agent
PBNA phenyl-β-naphthylamine
PBT polybutylene terephthalate
PC permeability coefficient; personal computer; plastic composite; plastic compounding; plastic-concrete; polycarbonate; printed circuit; process control; programmable circuit; programmable controller
PCB printed circuit board
pcf pounds per cubic foot
PCFC polychlorofluorocarbon
PDFM Plastics Distributors and Fabricators Magazine
PE plastic engineer; polyethylene (UK polythene); professional engineer
PEEK polyetheretherketone
PEI polyetherimide
PEK polyetherketone
PEN polyethylene naphthalate
PES polyether sulfone
PET polyethylene terephthalate
PETG polyethylene terephthalate glycol
PEX polyethylene crosslinked pipe
PF phenol formaldehyde
PFA perfluoroalkoxy (copolymer of tetrafluoroethylene and perfluorovinylethers)
PFBA polyperfluorobutyl acrylate
phr parts per hundred of rubber
PI polyimide
PIA Plastics Institute of America
PID proportional-integral-differential
PIM powder injection molding
PLASTEC Plastics Technical Evaluation Center (US Army)
PLC programmable logic controller
PMMA Plastics Molders and Manufacturers' Association (of SME); polymethyl methacrylate (acrylic)

PMMI Packaging Machinery Manufacturers' Institute
PO polyolefin
POE polyolefin elastomer
POM polyoxymethylene or polyacetal (acetal)
PP polypropylene
PPA polyphthalamide
ppb parts per billion
PPC polypropylene chlorinated
PPE polyphenylene ether
pph parts per hundred
ppm parts per million
PPO polyphenylene oxide
PPS polyphenylene sulfide
PPSF polyphenylsulfone
PPSU polyphenylene sulphone
PS polystyrene
PSB polystyrene butadiene rubber (GR-S, SBR)
PS-F polystyrene-foam
psf pounds per square foot
PSF polysulphone
psi pounds per square inch
psia pounds per square inch, absolute
psid pounds per square inch, differential
psig pounds per square inch, gauge (above atmospheric pressure)
PSU polysulfone
PTFE polytetrafluoroethylene (or TFE)
PUR polyurethane (also PU, UP)
P-V pressure-volume (also PV)
PVA polyvinyl alcohol
PVAC polyvinyl acetate
PVB polyvinyl butyral
PVC polyvinyl chloride
PVD physical vapor deposition
PVDA polyvinylidene acetate
PVdC polyvinylidene chloride
PVDF polyvinylidene fluoride
PVF polyvinyl fluoride
PVP polyvinyl pyrrolidone

PVT pressure-volume-temperature (also P-V-T or pvT)
PW *Plastics World* magazine
QA quality assurance
QC quality control
QMC quick mold change
QPL qualified products list
QSR quality system regulation
R Reynolds number; Rockwell (hardness)
rad Quantity of ionizing radiation that results in the absorption of 100 ergs of energy per gram of irradiated material.
radome radar dome
RAPRA Rubber and Plastics Research Association
RC Rockwell C (R_c)
RFI radio frequency interference
RH relative humidity
RIM reaction injection molding
RM rotational molding
RMA Rubber Manufacturers' Association
RMS root mean square
ROI return on investment
RP rapid prototyping; reinforced plastic
RPA Rapid Prototyping Association (of SME)
rpm revolutions per minute
RRIM reinforced reaction injection molding
RT rapid tooling; room temperature
RTM resin transfer molding
RTP reinforced thermoplastic
RTS reinforced thermoset
RTV room temperature vulcanization
RV recreational vehicle
Rx radiation curing
SAE Society of Automotive Engineers
SAMPE Society for the Advancement of Material and Process Engineering
SAN styrene acrylonitrile
SBR styrene-butadiene rubber
SCT soluble core technology
SDM standard deviation measurement
SES Standards Engineering Society
SF safety factor; short fiber; structural foam
s.g. specific gravity
SI International System of Units
SIC Standard Industrial Classification
SMC sheet molding compound
SMCAA Sheet Molding Compound Automotive Alliance
SME Society of Manufacturing Engineers
S-N stress-number of cycles
SN synthetic natural rubber
SNMP simple network management protocol
SPC statistical process control
SPE Society of the Plastics Engineers
SPI Society of the Plastics Industry
sPS syndiotactic polystyrene
sp. vol. specific volume
SRI Standards Research Institute (ASTM)
S-S stress-strain
STP Special Technical Publication (ASTM); standard temperature and pressure
t thickness
T temperature; time; torque (or T_t)
TAC triallylcyanurate
T/C thermocouple
TCM technical cost modeling
TD transverse direction
TDI toluene diisocyanate
TF thermoforming
TFS thermoform-fill-seal
T_g glass transition temperature
TGA thermogravimetric analysis
TGI thermogravimetric index
TIR tooling indicator runout
T-LCP thermotropic liquid crystal polymer
TMA thermomechanical analysis; Tooling and Manufacturing Association (formerly TDI); Toy Manufacturers of America
torr mm mercury (mmHg); unit of pressure equal to 1/760th of an atmosphere

TP thermoplastic
TPE thermoplastic elastomer
TPO thermoplastic olefin
TPU thermoplastic polyurethane
TPV thermoplastic vulcanizate
T$_s$ tensile strength; thermoset
TS twin screw
TSC thermal stress cracking
TSE thermoset elastomer
TX thixotropic
TXM thixotropic metal slurry molding
UA urea, unsaturated
UD unidirectional
UF urea formaldehyde
UHMWPE ultra-high molecular weight polyethylene (also PE-UHMW)
UL Underwriters Laboratories
UP unsaturated polyester (also TS polyester)
UPVC unplasticized polyvinyl chloride
UR urethane (also PUR, PU)
URP unreinforced plastic
UV ultraviolet
UVCA ultra-violet-light-curable-cyanoacrylate

V vacuum; velocity; volt
VA value analysis
VCM vinyl chloride monomer
VLDPE very low-density polyethylene
VOC volatile organic compound
vol% percentage by volume
w width
W watt
W/D weight-to-displacement volume (boat hull)
WIT water-assist injection molding technology
WMMA Wood Machinery Manufacturers of America
WP&RT World Plastics and Rubber Technology magazine
WPC wood-plastic composite
wt% percentage by weight
WVT water vapor transmission
XL cross-linked
XLPE cross-linked polyethylene
XPS expandable polystyrene
YPE yield point elongation
Z-twist twisting fiber direction

Acknowledgments

Undertaking the development through to the completion of the *Plastics Technology Handbook* required the assistance of key individuals and groups. The indispensable guidance and professionalism of our publisher, Joel Stein, and his team at Momentum Press was critical throughout this enormous project. The coeditors, Nick R. Schott, Professor Emeritus of the University of Massachusetts Lowell Plastics Engineering Department, and Marlene G. Rosato, President of Gander International Inc., were instrumental to the data, information, and analysis coordination of the eighteen chapters of the handbook. A special thank you is graciously extended to Napoleao Neto of Alphagraphics for the organization and layout of the numerous figure and table graphics central to the core handbook theme. Finally, a great debt is owed to the extensive technology resources of the Plastics Institute of America at the University of Massachusetts Lowell and its Executive Director, Professor Aldo M. Crugnola.

Dr. Donald V. Rosato, Coeditor and President, PlastiSource, Inc.

Preface

This book, as a two-volume set, offers a simplified, practical, and innovative approach to understanding the design and manufacture of products in the world of plastics. Its unique review will expand and enhance your knowledge of plastic technology by defining and focusing on past, current, and future technical trends. Plastics behavior is presented to enhance one's capability when fabricating products to meet performance requirements, reduce costs, and generally be profitable. Important aspects are also presented to help the reader gain understanding of the advantages of different materials and product shapes. The information provided is concise and comprehensive.

Prepared with the plastics technologist in mind, this book will be useful to many others. The practical and scientific information contained in this book is of value to both the novice, including trainees and students, and the most experienced fabricators, designers, and engineering personnel wishing to extend their knowledge and capability in plastics manufacturing including related parameters that influence the behavior and characteristics of plastics. The toolmaker (who makes molds, dies, etc.), fabricator, designer, plant manager, material supplier, equipment supplier, testing and quality control personnel, cost estimator, accountant, sales and marketing personnel, new venture type, buyer, vendor, educator/trainer, workshop leader, librarian, industry information provider, lawyer, and consultant can all benefit from this book. The intent is to provide a review of the many aspects of plastics that range from the elementary to the practical to the advanced and more theoretical approaches. People with different interests can focus on and interrelate across subjects in order to expand their knowledge within the world of plastics.

Over 20000 subjects covering useful pertinent information are reviewed in different chapters contained in the two volumes of this book, as summarized in the expanded table of contents and index. Subjects include reviews on materials, processes, product designs, and so on. From a pragmatic standpoint, any theoretical aspect that is presented has been prepared so that the practical person will understand it and put it to use. The theorist in turn will gain an insight into the practical limitations that exist in plastics as they exist in other materials such as steel, wood, and so on. There is no material that is "perfect." The two volumes of this book together contain 1800-plus figures and 1400-plus tables providing extensive details to supplement the different subjects.

In working with any material (plastics, metal, wood, etc.), it is important to know its behavior in order to maximize product performance relative to cost and efficiency. Examples of different plastic materials and associated products are reviewed with their behavior patterns. Applications span toys, medical devices, cars, boats, underwater devices, containers, springs, pipes, buildings, aircraft, and spacecraft. The reader's product to be designed or fabricated, or both, can be related directly or indirectly to products reviewed in this book. Important are behaviors associated with and interrelated with the many different plastics materials (thermoplastics [TPs], thermosets [TSs], elastomers, reinforced plastics) and the many fabricating processes (extrusion, injection molding, blow molding, forming, foaming, reaction injection molding, and rotational molding). They are presented so that the technical or nontechnical reader can readily understand the interrelationships of materials to processes.

This book has been prepared with the awareness that its usefulness will depend on its simplicity and its ability to provide essential information. An endless amount of data exists worldwide for the many plastic materials, which total about 35000 different types. Unfortunately, as with other materials, a single plastic material that will meet all performance requirements does not exist. However, more so than with any other materials, there is a plastic that can be used to meet practically any product requirement. Examples are provided of different plastic products relative to critical factors ranging from meeting performance requirements in different environments to reducing costs and targeting for zero defects. These reviews span products that are small to large and of shapes that are simple to complex. The data included provide examples that span what is commercially available. For instance, static physical properties (tensile, flexural, etc.), dynamic physical properties (creep, fatigue, impact, etc.), chemical properties, and so on, can range from near zero to extremely high values, with some having the highest of any material. These plastics can be applied in different environments ranging from below and on the earth's surface to outer space.

Pitfalls to be avoided are reviewed in this book. When qualified people recognize the potential problems, these problems can be designed around or eliminated so that they do not affect the product's performance. In this way, costly pitfalls that result in poor product performance or failure can be reduced or eliminated. Potential problems or failures are reviewed, with solutions also presented. This failure-and-solution review will enhance the intuitive skills of people new to plastics as well as those who are already working in plastics. Plastic materials have been produced worldwide over many years for use in the design and fabrication of all kinds of plastic products. To profitably and successfully meet high-quality, consistency, and long-life standards, all that is needed is to understand the behavior of plastics and to apply these behaviors properly.

Patents or trademarks may cover certain of the materials, products, or processes presented. They are discussed for information purposes only and no authorization to use these patents or trademarks is given or implied. Likewise, the use of general descriptive names, proprietary names, trade names, commercial designations, and so on does not in any way imply that they may be used freely. While the information presented represents useful information that can be studied or analyzed and is believed to be true and accurate, neither the authors, contributors, reviewers, nor the publisher can accept any legal responsibility for any errors, omissions, inaccuracies, or other factors. Information is provided without warranty of any kind. No representation as to accuracy, usability, or results should be inferred.

Preparation for this book drew on information from participating industry personnel, global industry and trade associations, and the authors' worldwide personal, industrial, and teaching experiences.

DON & MARLENE ROSATO AND NICK SCHOTT, 2011

About the Authors

Dr. Donald V. Rosato, president of PlastiSource Inc., a prototype manufacturing, technology development, and marketing advisory firm in Massachusetts, United States, is internationally recognized as a leader in plastics technology, business, and marketing. He has extensive technical, marketing, and plastics industry business experience ranging from laboratory testing to production to marketing, having worked for Northrop Grumman, Owens-Illinois, DuPont/Conoco, Hoechst Celanese/Ticona, and Borg Warner/G.E. Plastics. He has developed numerous polymer-related patents and is a participating member of many trade and industry groups. Relying on his unrivaled knowledge of the industry and high-level international contacts, Dr. Rosato is also uniquely positioned to provide an expert, inside view of a range of advanced plastics materials, processes, and applications through a series of seminars and webinars. Among his many accolades, Dr. Rosato has been named Engineer of the Year by the Society of Plastics Engineers. Dr. Rosato has written extensively, authoring or editing numerous papers, including articles published in the *Encyclopedia of Polymer Science and Engineering*, and major books, including the *Concise Encyclopedia of Plastics*, *Injection Molding Handbook 3rd ed.*, *Plastic Product Material and Process Selection Handbook*, *Designing with Plastics and Advanced Composites*, and *Plastics Institute of America Plastics Engineering, Manufacturing, and Data Handbook*. Dr. Rosato holds a BS in chemistry from Boston College, an MBA from Northeastern University, an MS in plastics engineering from the University of Massachusetts Lowell, and a PhD in business administration from the University of California, Berkeley.

Marlene G. Rosato, with stints in France, China, and South Korea, has comprehensive international plastics and elastomer business experience in technical support, plant start-up and troubleshooting, manufacturing and engineering management, and business development and strategic planning with Bayer/Polysar and DuPont. She also does extensive international technical, manufacturing, and management consulting as president of Gander International Inc. She also has an extensive writing background authoring or editing numerous papers and major books, including the *Concise Encyclopedia of Plastics*, *Injection Molding Handbook 3rd ed.*, and the *Plastics Institute of America Plastics Engineering, Manufacturing and Data Handbook*. A senior member of the Canadian Society of Chemical Engineering and the Association of Professional Engineers of Canada, Ms. Rosato is a licensed professional engineer of Ontario, Canada. She received a Bachelor of Applied Science in chemical engineering from the University of British Columbia with continuing education at McGill University in Quebec, Queens University and the University of Western Ontario, both in Ontario, and also has extensive executive management training.

Emeritus Professor Nick Schott, a long-time member of the world-renowned University of Massachusetts Lowell Plastics Engineering Department faculty, served as its department head for a quarter of a century. Additionally, he founded the Institute for Plastics Innovation, a research consortium affiliated with the university that conducts research related to plastics manufacturing, with a current emphasis on bioplastics, and served as its director from 1989 to 1994. Dr. Schott has received numerous plastics industry accolades from the SPE, SPI, PPA, PIA, as well as other global industry associations and is renowned for the depth of his plastics technology experience, particularly in processing-related areas. Moreover, he is a quite prolific and requested industry presenter, author, patent holder, and product/process developer. In addition, he has extensive and continuing academic responsibilities at the undergraduate to postdoctoral levels. Among America's internationally recognized plastics professors, Dr. Nick R. Schott most certainly heads everyone's list not only within the 2500 plus global UMASS Lowell Plastics Engineering alumni family, which he has helped grow, but also in broad global plastics and industrial circles. Professor Schott holds a BS in chemical engineering from UC Berkeley, and an MS and PhD from the University of Arizona.

CHAPTER 18

Auxiliary and Secondary Equipment

INTRODUCTION

An important sector of the plastics industry is the machinery and equipment sector (Table 18.1). Many different fabricating lines are used to provide the millions of plastic products used worldwide. These lines have primary, auxiliary, and secondary equipment. Primary equipment refers to the machine that starts fabricating a product, such as an injection molding machine, extruder, blow molder, thermoformer, and so on. This equipment has been reviewed in chapters 4 to 17. Auxiliary equipment (AE) and secondary equipment (SE) support the primary equipment. This type of equipment is required in order to produce products that fit into the overall manufacturing cycle. There are many different types of equipment that support nonautomated and automated upstream and downstream production in-line or off-line systems that maximize the overall processing efficiency of productivity and reduce operating cost (Table 18.2). Examples of this equipment are shown in Figures 18.1 to 18.10.

Even though modern fabricating lines, with their ingenious dies and molds and microprocessor control technology, are in principle suited to perform flexible tasks, it nevertheless takes a whole series of peripheral AE or SE or both to guarantee the necessary degree of flexibility. Examples of this action are provided in Tables 18.3 and 18.4.

Throughout this book, auxiliary and SE have been included in reviews of different primary equipment. An example is the review in chapter 5 on the many different control sensors that include level sensors for plastic material inventory, transferring plastic to fabricating equipment, controlling the primary and secondary fabricating equipment, controlling the fabricated product, packaging product, and shipment of product. This chapter will provide details and summations on the equipment make up a quarter-million-dollar business in the United States (570).

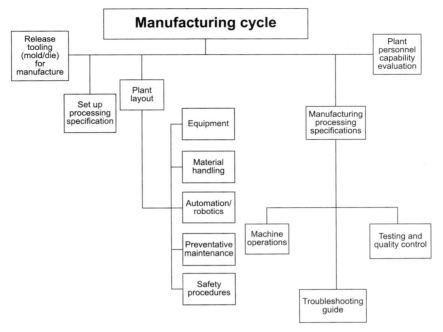

Table 18.1 Example of manufacturing cycle that includes equipment

AE can sometimes cost more than the primary equipment. Requirements for the AE to be used are based on what is reliable and what is required in the production line. It is important to properly determine requirements and ensure that the AE interface into the line (size, capacity, speed, etc.); otherwise many costly problems can develop. They have become more energy efficient, reliable, and cost effective. The application of microprocessor- and computer-compatible controls that can communicate with the line (train) results in pinpoint control of the line. A set of rules have been developed and used by equipment manufacturers; these rules help govern the communication protocol and transfer of data between primary and AE.

Ideally, fabricating thermoplastic (TP) or thermoset (TS) plastic products will be finished as processed. For example, almost any type of texture, surface finish, or insert can be fabricated into the product, as can almost any geometric shape, hole, or projection. There are situations, however, where it is not possible, practical, or economical to have every feature in the finished product. Typical examples where machining might be required are certain undercuts, complicated side coring, or places where parting line or weld line irregularity is unacceptable. Another common machining and finishing operation for plastics is removal of the remnant of the flash, sprue, or gate (or all three) if it is in an area of the part where appearance is important or in an area where tolerance is critical.

COMPANY	Accumulators	Blenders	Bonding	Chillers	Computers	Decorating	Dryers	Fabrication	Granulators	Machine Components	Materials	Measuring	Molds Dies	Parts	Pelletizers	Scrap	Screen Changer	Takeoff	SPC	Testing
Adams Engineers And Equipment Inc Tyler, TX (903) 561-8835		•		•			•		•		•			•			•			
AEC Inc, Wood Dale, IL (630) 595-1060		•		•			•		•		•			•						
Agr* TopWave LLC, Butler, PA (724) 482-2163															•	•				
Ai-Be Industries, Paramount, CA (562) 272-4646	•																			
Air Logic Power Systems, Inc, Milwaukee, WI (414) 671-3332		•		•																
American LEWA, Holliston, ME (508) 429-7403												•								
American MSI Corporation, Moorpark, CA (805) 523-9593						•						•								
American Screw & Barrel, Inc, Gardner, MA (978) 630-1300										•		•	•							
Anderson American Precision Inc, Cocoa, FL (321) 637-0728										•										
Anter Corporation, Pittsburgh, PA (412) 795-6410												•								•
Apex Machine Company, Fort Lauderdale, FL (954) 565-2739						•														
Applied Robotics, Glenville, NY (518) 565-2739			•											•						
Atlas Vac Division, Planet Products, Cincinnati, OH (513) 984-5544			•			•														
ATS RheoSystems, Bordentown, NJ (609) 298-2522												•								•
Automated Assemblies Corporation, Clinton, MA (978)368-8914						•		•						•						
Automatic Filters, Inc, Los Angeles, CA (310) 839-2828			•	•																
Automation Technology, St Louis, MO (314) 567-3100					•		•				•			•						
B A Die Mold, Inc, Aurora IL (630) 978-4747													•							
Battenfeld Gloucester Engineering Co Inc, Gloucester, MA (978) 281-1800	•						•	•			•									•
BF Perkins Machinery, Rochester, NY (585) 292-5280							•				•			•				•		
Biasdel Enterprises, Inc, Greensburg, IN (812) 663-3212											•									
Bohlin Instruments, East Brunswick, NJ (732) 254-7742												•								•
Bolz-Summix, Pennsauke, NJ (856) 317-9960		•																		
Briquetting Systems, Vancouver, BC (604) 818-0287								•							•					
Bry-Air System, Sunbury, OH (740) 965-2974				•			•													
Budget Molders Supply, Macedonia, OH (216) 367-7050				•									•	•						
Burger & Brown Engineering Inc, Olathe, KS (913) 764-3518				•			•													
Casso-Solar Corporation, Pamona, NY (800) 988-4455						•		•												
CBC (America Corp), Commack, NY (800) 422-6707											•		•							
Chem-Pak Inc, Martinsburg, WV (304) 262-1880			•																	
Cleveland Process Corporation - CLEPCO, Homestead, FL (800) 241-0412											•		•	•				•		
CMT Materials, Inc, Attleboro, MA (508) 226-3901											•									
Columbian TecTank, Parsons, KS (620) 421-0200											•									
Conair, Pittsburgh, PA (412) 312-6000		•		•			•				•			•	•			•		
Consolidated Polymer Technologies, Clearwater, FL (800) 541-6880								•			•									
Contech, Goddard, KS (316) 722-6907	•							•												
Controls Southeast Inc, Pineville, NC (704) 644-5000											•		•							
Coperion Corporations, Ramsey, NJ (201) 327-6300											•			•	•			•		
Copper and Brass Sales, Detroit, MI (313) 566-7425											•		•	•						

Table 18.2 SPE auxiliaries buyer's guide (courtesy of SPE)

AUXILIARY AND SECONDARY EQUIPMENT   741

Table 18.2 SPE auxiliaries buyer's guide (courtesy of SPE) (continued)

Table 18.2 SPE auxiliaries buyer's guide (courtesy of SPE) *(continued)*

Table 18.2 SPE auxiliaries buyer's guide (courtesy of SPE) *(continued)*

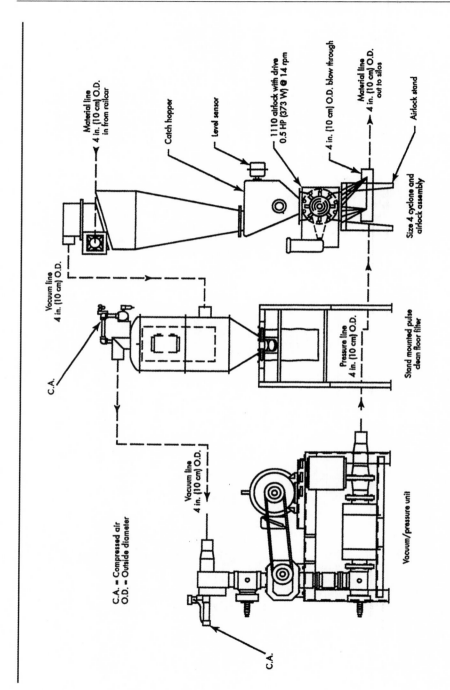

Figure 18.1 Example of AE required for plastics going from a railcar to a silo.

Figure 18.2 Closeup view of a piping system to and from silos, with each having a capacity of 2000 lb.

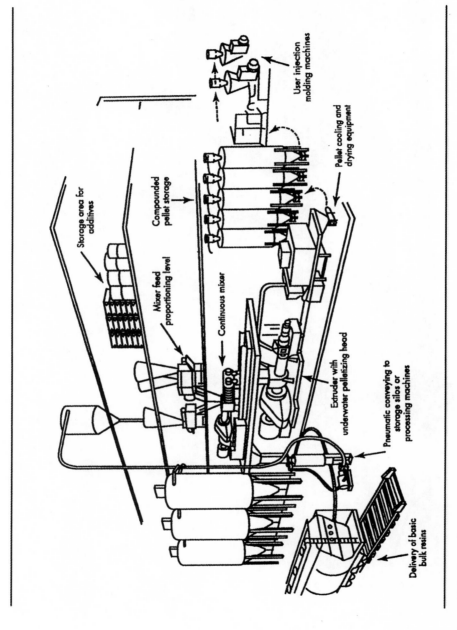

Figure 18.3   Examples of plant layout with extrusion and injection molding primary and AE.

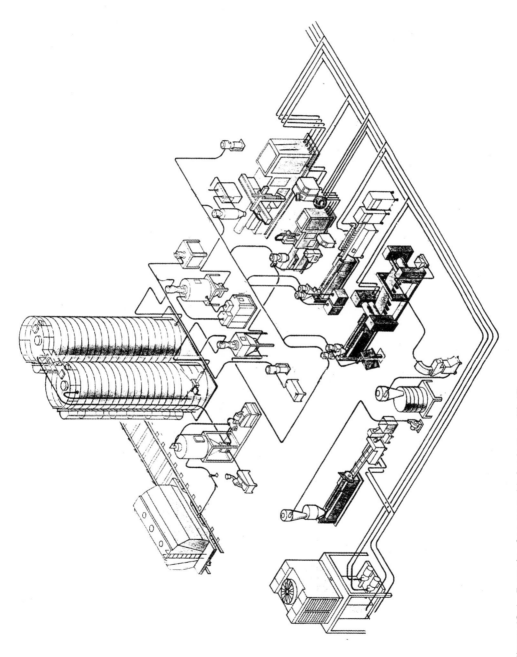

Figure 18.4 Example of an extrusion laminator with AE.

Figure 18.5 Example of a blow-molding extruder with AE (rolls, turret winder, etc.).

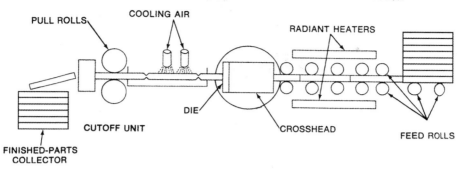

Figure 18.6 Example of an extruder coater with AE.

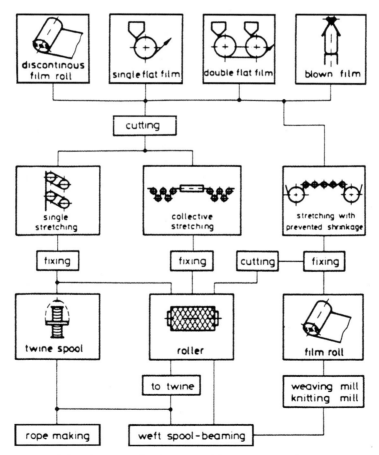

Figure 18.7 Example of plant layout with injection molding primary and AE.

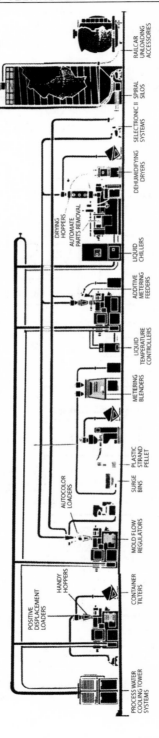

Figure 18.8 Example of extruded products requiring AE.

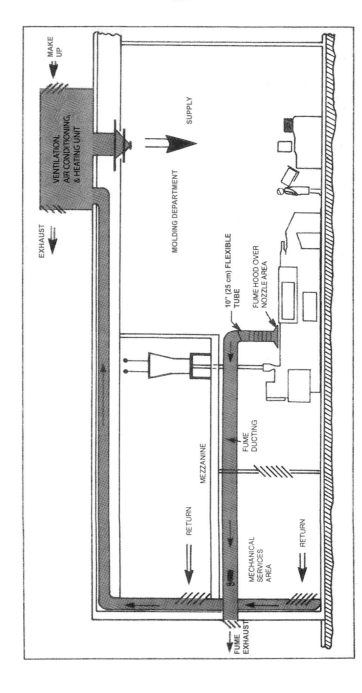

Figure 18.9 Example of ventilation AE used with an injection molding machine (courtesy of Husky Injection Molding Systems Inc.).

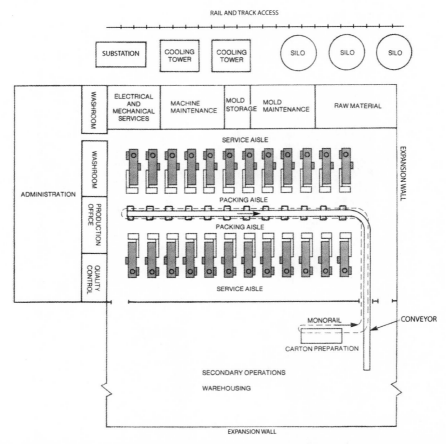

Figure 18.10 Examples of material handling AE used with an injection molding machine (courtesy of Husky Injection Molding Systems Inc.).

These secondary operations can occur in-line or off-line. They include any one operation or a combination of operations, such as machining, annealing (to relieve or remove residual stresses and strains), postcuring (to improve performance), plating, joining and assembling (adhesive, ultrasonic welding, vibration welding, heat welding, etc.), cutting, finishing, polishing, labeling, decorating, and printing. The type of operation to be used depends on the type of plastic used. For example, in decorating or bonding, certain plastics can be easily handled while others require special surface treatments to produce acceptable products.

Heat is applied to seal and join pliable plastics sheet (less than 50 mils thick); heat sealing is limited to use on TP materials. The heat may be provided by thermal, electrical, or sonic energy. A wide variety of heat-sealing systems are available.

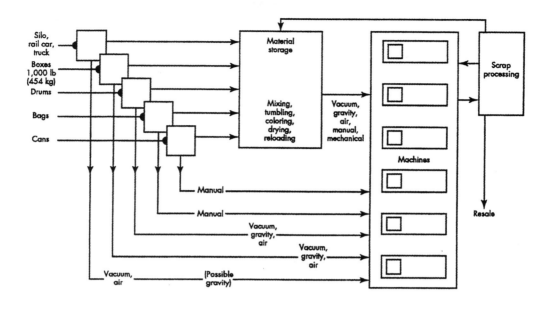

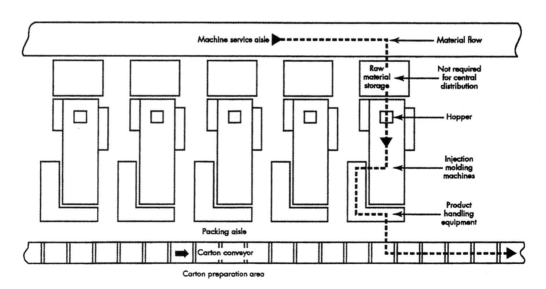

Figure 18.10 Examples of material handling AE used with an injection molding machine (courtesy of Husky Injection Molding Systems Inc.) *(continued)*.

1. up stream material supply systems (dryers, blenders, etc.),
2. mold or die transport facilities,
3. mold or die preheating banks,
4. mold or die changing devices that includes rapid clamping and coupling equipment,
5. plasticizer cylinder changing devices (screws, barriers, etc.),
6. molded or extruded product handling equipment, particularly robots with interchangeable arms allowing adaptation to various types of production,
7. transport systems for finished products and handling equipment to pass products on to subsequent production stages.
8. and more.

Table 18.3 Introduction to auxiliary and SE performances

Plastic sections, which are too thick to be heat sealed, may usually be welded. There are three major methods of sealing in commercial use: heat, solvent, and ultrasonic. In general, these methods are limited to use with TP materials. These welding techniques have done much to lower the total cost of using plastics in construction and other industries.

In addition to the various welding techniques, adhesives may join plastic parts. Both TP and TS resins may be bonded and parts made of different resins are often treated in this manner. There is a wide range of suitable adhesive materials, including various monomers, solvents, and epoxies, that are in general commercial use. The choice of material will be a function of the plastic materials to be joined and the environmental and end-use conditions to which the finished part will be subjected.

The increasing use of plastics as construction materials has led to a renewed interest in decorative finishes for plastic products. There is a wide variety of secondary operations that can be used for adding decoration to molded parts. Progress is also being made in providing decorative surfaces in the mold itself. The first use of this is in wood like panels for wall decoration and furniture parts such as cabinet doors.

Plastics may be printed upon, painted by a variety of processes, wood-grained through a printing process, electroplated, metallized, and hot stamped with gold or silver leaf. Plastic film and sheeting are generally printed or embossed to create decorative surfaces. Printing is also used in the mass production of such plastic articles as labels, signs, and advertising displays.

Adhesive applicator	Process control for individual or all equipment
Bonding	
Chemical etching	
Cutting	Pulverizing/grinding
Decorating	Quick mold change
Die cutting	Recycling system
Dryer	Robotic handling
Dust-recovery	Router
Flash removal	Saw
Freezer/cooler	Screen changer
Granulator	Screw/barrel backup
Heater	Sensor/monitor control
Joining	Software
Knitting	Solvent recovery
Labeling	Solvent treatment
Leak detector	Testing/instrumentation
Machining	Trimming
Material handling	Vacuum debulking
Metal treating	Vacuum storage
Metering/feeding material	Water-jet cutting
	Welding
Mold extractor	Others
Mold heat/chiller control Oven	
Pelletizer/dicer	
Plating	
Polishing	
Printing/making	

Table 18.4 Examples of auxiliary and SE

There has been increasing interest in the process of electroplating plastics. Plating can produce chromelike, brass, silver, gold, or copper surfaces in both smooth and textured forms. There are several systems for plating plastic materials available commercially. Certain plastics, such as electroplated acrylonitrile butadiene styrene (ABS), can have surfaces chemically treated to promote bonding of the metals in subsequent steps.

This action eliminates the need for a costly mechanical "roughening" process that most other materials require. The depositing of a metal surface on plastic parts increases the environmental resistance of the part, its mechanical properties, and its appearance. For example, a plated ABS part (total thickness of plate 0.015 in) exhibited a 16% increase in tensile strength, a 100% increase in tensile modulus, a 200% increase in flexural modulus, a 30% increase in Izod impact strength, and a

12% increase in deflection temperature. Tests on outdoor-aged samples showed complete retention of physical properties after six months.

It is possible for plated plastics to corrode if the metal coating is not properly applied or if it is damaged in such a way as to allow electrolytic interaction in the plating layers. On the other hand, the plastic substrate will not corrode itself, nor will it contribute to further corrosion of the plating layers. In general, plated plastics will fare better than metals when exposed to corrosive environments. No one, however, suggests that plastics will replace metals or plated metals in all uses, or even in a majority of uses, since the physical and mechanical properties of the materials and the end-use requirements may be overriding influences.

MATERIAL/PRODUCT HANDLING

Design of the raw material handling system has a major impact on the plant's manufacturing costs and housekeeping. It is based on the different materials used, annual volume of each material, number of different colors, production run lengths, and so on. A properly designed pneumatic system generates plastic velocities of 5000 ft/min (1500 m/min).

Material handling can start with the truck or railcar delivery of bulk plastic to the plant. Trucks typically carry 1250 ft^3 (35 m^3) of material. Most often the truck has a positive displacement-pumping unit, or the user supplies a pressure system to the silos. Railcars can store up to 5200 ft^3 (147 m^3) in four or five compartments, with the user providing unloading systems to the silos. Unloading costs are largely determined by the throughput required.

Plastics may be supplied in different quantities. There are drums (from 15 lb [11 kg]), bags (50 lb [23 kg]), gaylords (cardboard boxes, usually lined with plastic sheets, that hold 1000 lb [454 kg]), as well as bulk fabric sacks or bags (also called super sacks, super bags, or jumbo bags, holding 2000 lb [907 kg]). It may be appropriate to purchase plastics in one of these containers rather than in bulk because of low-volume use, cost, a moisture situation, or other factors. To move materials from these containers, vacuum-tube conveyors (Fig. 18.11), dumper or pressure unloaders, fork truck hoists, and so on, may be used. Plastic storage box containers may also be purchased instead of bags or drums. Box sizes and weights vary but nearly always conform to a standard pallet on which they are shipped and moved around inside the plant.

The bulk density of the plastic influences solids conveying and processing. The *bulk density* is the weight of a unit volume of the material including air voids. The actual *material density* is defined as the weight of a unit volume of the plastic excluding air voids.

If the bulk density is more than 50% of the actual density, the bulk material likely will be reasonably easy to convey in a material-handling system and through a plasticator (chapter 3). In this case, the screw channel depth in the plasticator's feed section does not have to be too large—between 0.1 diameter and 0.2 diameter—and the compression ratio can be at the low end of the range, from about 1 to 3.

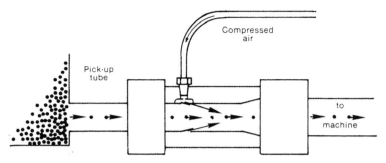

Figure 18.11 Example of a pneumatic vacuum venturi flow system.

If the bulk density is less than 50% of the actual density, then solids-conveying problems are likely to occur. With these materials, the deep-feed section has to be rather deep to obtain sufficient solids conveying. As a result, the compression ratio will be at the high end of the range, from about 3 to 5.

When the bulk density becomes less than 30% of the actual density, a conventional plasticator usually cannot handle the bulk material. Such materials may require special feeding devices, such as crammer feeders and special extruder designs, such as a large-diameter feed section tapering down to a smaller diameter metering section.

Material-Handling System

Different methods are used to move plastic. They range from manual methods to fully automated methods, for either raw material or processed parts. The methods are used in automatic bulk systems, in-line granulators, parts-removal robots, conveyors, stackers or orienters, and so on. The equipment chosen must match the productivity requirements.

When conveying plastics a properly designed system is to take the shortest distance. Shortest distance between two points is a straight line. The maximum conveying distance is usually 800 equivalent ft (244 m). A gradual upward slope is never better than a vertical lift. When the plastic passes through a 45° or 60° elbow, it ricochets back and forth, creating turbulence that destroys its momentum.

With vacuum or pressure, the conveying action provides double the conveying rates of vacuum alone. Plastic lines are not recommended for conveying lines since static electricity will be generated and will interfere with the movement of plastics. There is a rather simple and useful test to determine if material is going to be difficult to convey. Take a handful of the plastic and squeeze it firmly. Upon opening your hand, if the lines in your palm are filled with fines, the plastic will be difficult to convey.

Fines are very small particles, usually under 200 mesh, accompanying larger forms of powders. When plastics are extruded and pelletized, varying amounts of oversized pellets and strands are produced, along with fines. When the plastics are dewatered (or dried) or pneumatically conveyed, more fines, fluff, and streamers may be generated. They can develop when plastics are granulated. Usually they are detrimental during processing so they are removed, or action is taken to eliminate the problem during pelletizing, scrap grinding, and so on.

In addition to conveying plastic, there is a wide variety of tasks in warehousing, such as handling raw materials, additives, AE, spare parts, molds, dies, tools, processed plastic parts, and so on. They require proper handling and storage procedures that are logged consistently. Various systems, such as the unit warehouse that makes use of pallets, cages, and similar equipment, are used successfully. They employ a certain organizational scheme for integrating order picking and transportation. The system is perfected by integration of the inward and outward flow (input-output matrix) of goods, the factory administration, process control, quality control, and so on.

Pneumatic conveying

The handling and conveying of bulk material involves a combination of theory and experience. Almost any substance (pellets, granules, powders, etc.) can be conveyed pneumatically (i.e., with air) increasing the economic advantage of materials purchase, storage, and use throughout the plant (such as in filling fabricating machine hoppers). Velocity, which is defined as the rate of motion or speed, is the key to transporting materials. Pipeline conveyors are commonly referred to as dilute-phase systems (i.e., conveying a small volume of material using a large volume of air).

The physics principles that play an important role in pneumatic conveying are gravity, the pressure differential (the force resulting from a difference in pressure that is used to initiate the movement of air and material), inertia, shear (between adjacent particles), and elasticity (the intrinsic tendency of a compressible gas to expand and flow from a high-pressure region to a low-pressure region). The basic laws of physics affecting fluid flow are the laws of conservation of matter and energy, the perfect gas properties, and the gas laws based on Boyle's law, Gay-Lussac's law, Charles's law, and the combined gas law. Detailed equations are available in handbooks on physical principles. These formulas explain the flow and work capabilities of air in a pneumatic system. However, these characteristics of air are only one factor in the art of pneumatic conveying. The other factor concerns the actual material that is to be conveyed pneumatically.

A pneumatic conveying system can move materials from the storage area to the processing machine automatically and with little risk of the contamination of the materials that commonly occurs when materials are moved to the processing machine manually. The major problem with the pneumatic conveying of materials is the absence of a standard set of formulas for calculating the equipment sizing and flow characteristics of the material to be conveyed. Standard formulas do not exist because of the vast variety of plastic materials, additives, and their combinations. Most available information on the sizing of material-handling systems is based on experiments conducted by equipment manufacturers and the observation of successful conveying systems currently in operation.

Pipeline conveyors are involve the conveying of a small volume of material by a large volume of air. This is put into a ratio commonly called the material-to-air ratio and classified in pounds (kilograms). The counterpart to the dilute phase is the dense phase. Dense-phase conveying is described as a pound of air moving material equal to weight through the conveying tubes. In other words, material is conveyed in a dense-phase state that will have a low material-to-air ratio. Compactable powders are common materials that are conveyed in a dense state.

Pneumatic conveying is a branch of hydraulics power and not a separate form of power transmission. The explanation of the term *pneumatics* is based on the fact that every liquid has a temperature and a pressure point where it becomes a gas. A common example of this is water. When water is brought to a boiling point at 100°C (212°F) with atmospheric pressure at 14.696 psi, the water turns to steam, which is water in a gaseous state. The boiling point of liquids varies with the amount of pressure and temperature in a controlled situation at 32°F (0°C), which is the normal freezing point of water. However, with a pressure reading of 0.0885 psia, the reduction of atmospheric pressure will actually force the water to boil at that low temperature.

The behavior of air is the same as the behavior of water. Air has a temperature and pressure point at which it becomes a liquid substance. Liquids and gases have distinct characteristics that separate the two substances. Gases generally can be compressed easily, which makes gases an excellent medium when a substance is needed to perform a task in a limited space, such as a pipe. Air and other gases can be compressed and stored in containers to eventually be used to perform work. A gas can be compressed to an extreme pressure, where the temperature can be raised to force the substance to condense back into a liquid and be stored in a tank.

With the tank opened and the substance in liquid form released into the atmosphere, the liquid will return to a gaseous state because the pressure and temperature will immediately be lowered. Once a substance reaches a liquid state, it generally cannot be compressed into a smaller mass. The characteristics of a substance can be predicted and measured more easily when the substance is in a liquid state. It is this predictability of substances in a liquid that provides the basis for formulas used in pneumatic conveying.

Conveying rate and distance

Empirical formulas for pneumatic conveying have not been established because of the different behaviors and different shapes and sizes of materials that can be conveyed by air. Manufacturers of conveying and storage equipment have established simple sizing charts based on the most commonly used materials in the plastics industry. This work is based on the fact that most plastic materials have an average bulk density of 35 lb/ft^3. Polystyrene has a bulk density of 35 lb/ft^3 (561 kg/m^3) in both pellet and powder form.

Various tests have been conducted on polystyrene pellets and powders to determine which conveying rates, distances, line sizes, and air pressures are appropriate to convey these two substances. It was discovered that most materials can be conveyed between 6 and 10 psi gauge pressure. From this information, graphs were designed for easier sizing of a pneumatic system. Figures 18.12 to 18.16 show vacuum and pressure systems that contain a total lift and run conveying loop with four elbows included.

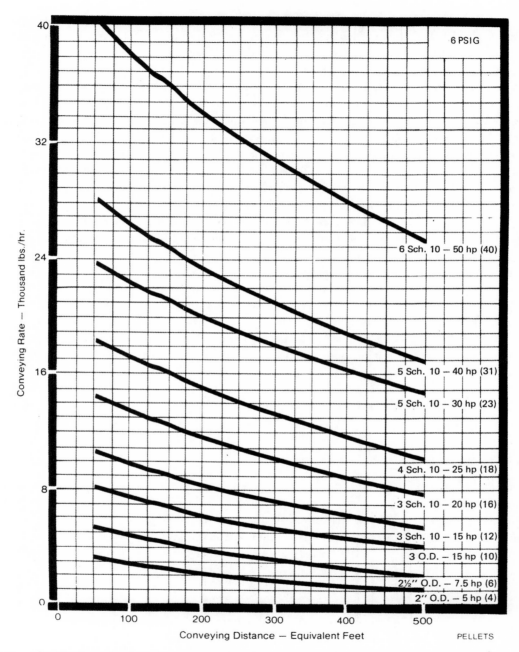

Figure 18.12 Example of continuous pressure pellets with rates based on polystyrene at 35 lb/ft^3 (560 kg/m^3).

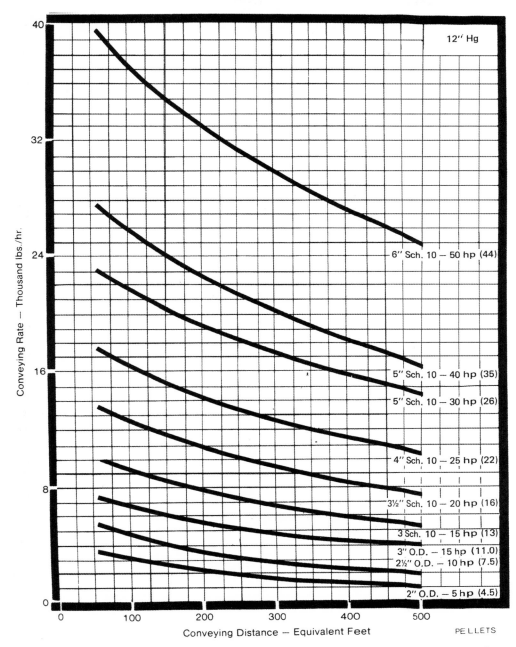

Figure 18.13 Example of continuous vacuum pellets with rates based on polystyrene at 35 lb/ft^3 (560 kg/m^3).

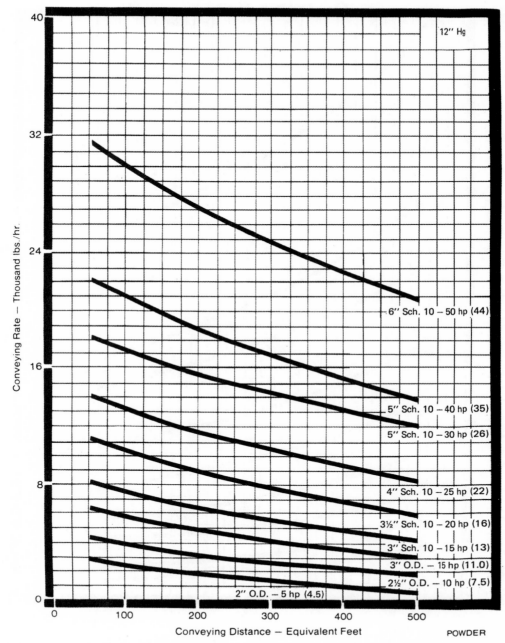

Figure 18.14 Example of continuous vacuum powder with rates based on polyvinyl chloride (PVC) at 35 lb/ft^3 (560 kg/m^3).

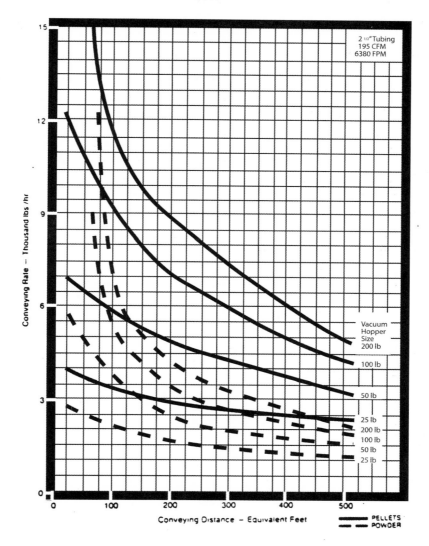

Figure 18.15 Example of a 10 hp vacuum system conveying polystyrene at 35 lb/ft^3 (560 kg/m^3).

There are some rules of thumb that can be included in this section on graphs and sizing. To make the sizing procedure less complicated, when tubing that is not straight is encountered in the system, one converts the pressure drop of the bend, flex-hose, and so on into an equivalent length of straight tubing.

When a graph is used, a 10% conversion factor is added to the conveying rate for each elbow under the number of four that is not used. The 10% should be subtracted for every elbow up to

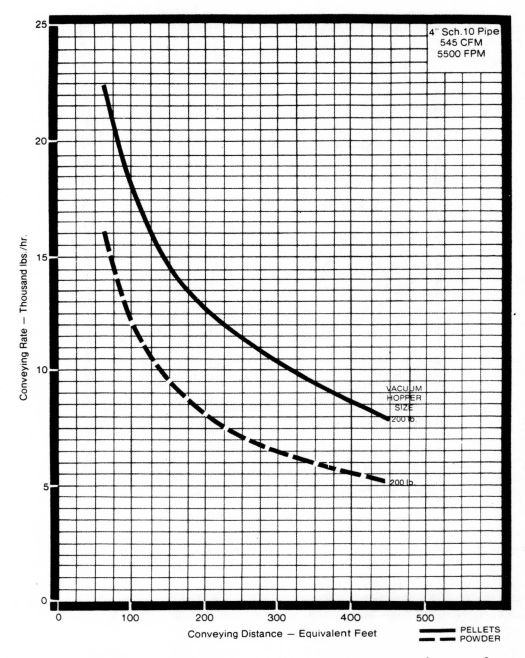

Figure 18.16 Example of a 25 hp vacuum system conveying polystyrene at 35 lb/ft^3 (560 kg/m^3).

six total elbows. If more than six total elbows are used, the manufacturer should be contacted for specific conversion factors relating to a particular material. Other rules of thumb exist.

Linear length of piping is adjusted for the direction of flow. For example, if the flow is vertical, one unit of length is equal to two units of equivalent length. A continuous vacuum/pressure system provides the ability to convey material at nearly the rate required (15000 to 30000 lb [6804 to 13608 kg] per hour is typical) over almost any distance required (<400 ft [<122 m] on the vacuum side, 1200 ft [366 m] on the pressure side is typical). It is sometimes cost effective to purchase two lower throughput systems that can unload two cars simultaneously rather than one high-volume system.

With automatic delivery from silos into the plant, all plastic-handling lines are kept as short as possible. There is no reason for lines to conform to the right angles of the walls; they should follow a straight line from the plastic's source to where it has to be delivered. There are graphs from suppliers of handling systems that show the relationship between the length of conveyor lines and power requirements. The graphs also show the horsepower (hp) required, based on different factors such as the length and diameter of the delivery pipe, the position of the pipe, the type of plastic being conveyed, the size of the hopper at the machine, and the rate of flow deliverable. Table 18.5 provides information on cost savings when using energy-efficient electric motors.

Operating		8 hrs/day, 5 days/wk			16 hrs/day, 5 days/wk			24 hrs/day, 7 days/wk		
HP	RPM	$/kWh 0.05	$/kWh 0.07	$/kWh 0.10	$/kWh 0.05	$/kWh 0.07	$/kWh 0.10	$/kWh 0.05	$/kWh 0.07	$/kWh 0.10
10	3,600	51	72	103	103	144	205	215	301	430
	1,800	42	59	84	84	117	167	175	245	350
	1,200	48	67	96	96	134	192	201	281	401
20	3,600	76	106	152	152	213	304	318	445	635
	1,800	58	81	115	115	162	231	241	338	483
	1,200	59	82	118	118	165	236	246	345	493
50	3,600	183	256	366	366	512	732	766	1,072	1,531
	1,800	97	136	195	195	273	389	407	570	814
	1,200	119	167	238	238	334	477	499	698	997
100	3,600	303	424	605	605	847	1,211	1,266	1,772	2,532
	1,800	227	318	455	455	637	909	951	1,331	1,902
	1,200	264	370	529	529	740	1.058	1,106	1,548	2,212

Table 18.5 Estimated annual savings for energy-efficient electric motors (Electrical Apparatus Service Association)

Plastic property influence

Various properties and characteristics of materials used in the plastics industry that can be conveyed pneumatically affect the sizing and design of the conveying system. For example, the specific gravity is an aid in determining how much airflow is needed to lift a particle in an airstream. Particle size is also a consideration in pneumatic conveying systems. The material has to be tested to determine the amount of fines and dust that may be contained in the material. This will help determine the type of airflow in a system and whether it is a vacuum or pressure system, along with the type of filters that will be utilized in the system. Particle size is measured with sieves that are made to standards set by the American Society for Testing Materials of the US Standards Institute (chapter 13).

A common method of testing for the particle size and range would be to place the material in a sieve with a large mesh opening and shake the material until the small particles pass through the screen. This procedure is repeated, with screens of a smaller mesh size, until all the particulate is separated.

If the material is very tacky, it may not be suitable to be conveyed pneumatically. Tacky material may smear against conveying pipes and cause a buildup of material that will eventually clog a line. A simple test for material tackiness would be to take some material into your hand and squeeze it into a compact ball. If the material sticks together, it is classified as a tacky material.

The melting point of a material should be determined. There are plastic materials that melt at low temperatures. If these materials are conveyed at a faster rate than necessary, they may slide against the walls of the conveying tubes and heat up by friction, which in turn will cause them to begin to melt, producing what is called "angel hair." This commonly takes place in a bend of a conveying tube due to the centrifugal force that is placed on the pellets or other forms of plastic materials, forcing them to slide along the outer periphery of the tube.

The melting plastic running along the wall, leaving a thin trail of plastic along the tube wall, causes angel hair. This thin plastic will partially peel away from the wall as the pellet moves back toward the center of the air stream, leaving what appears to be a fine hair. If enough of this occurs with other pellets in a particular area in a system, the angel hair will clog the system, thus preventing material from flowing through the system.

Abrasive materials may cause the conveying tubes to wear through quickly. Abrasive materials may have to be conveyed through resistant material or at a lower rate than other materials if at all possible.

Odors and the toxicity of materials should be considered when developing a conveying system. These two related factors are not very common in the plastics industry. However, these characteristics could be a common element when dealing with other chemicals that may be related to the plastics industry. These elements have an effect on the type of conveying system and filtration system that should be incorporated into a plant.

Very few plastics have a corrosive characteristic that may contain acids and erode tubing. An acid-content test can be conducted by determining pH factor. A pH of 7 is neutral. Any reading below 7 is an indication of acid. A pH reading above 7 indicates that the material is alkaline.

Powdered materials with strong acid indications will have to be conveyed through special pneumatic systems in order to prevent any corrosion from taking place within the system.

Plastics flow is influenced by the plastic's ability to be saturated by air in a free-flowing state. If a material can be continuously saturated by air in a free-flowing state, the material is aerated. Should a material clump together and block the airflow, the material is deaerated. Tests can be performed on materials to determine aeration characteristics. The material can be placed in a container with a lid on it. The container is shaken for a few moments. If the volume of material appears to have increased and is taking a long time to settle to the bottom of the container, the material is an aerated material. If the material settles to the bottom of the container quickly, it is a deaerated material.

Another aeration test uses the angle of repose. The material is put on a horizontal plane, and that plane is lifted at an angle until the material starts to flow; that is the angle of repose. If the material starts to flow at a low angle, the material is said to be free-flowing or aerated material. If the material flows at a high angle, the material is said to be deaerated or hard-flowing material. The angle of repose not only helps with the sizing of pneumatic systems, but it also aids one in choosing the proper storage system equipment.

Control feeding device

Control feeding devices to the hopper of primary equipment (injection molding, extrusion, etc.) is important to provide products that meet performance requirements at the lowest cost. Equipment manufacturers have increased the feeding accuracy using different devices such as microprocessor blender/mixer controllers. Materials are being reduced in size with more uniformity to significantly improve uniformity in melt. Processors can use blenders and other devices mounted on hoppers that target for precise and even distribution of materials (Figs. 18.17 to 18.28).

Hooper feeding

Hoppers are receptacles on the machines that direct the plastic materials (pellets, granules, flakes, etc.) being fed into the plasticators. The hopper can be fitted with devices to perform different functions. For example, they can be fitted with a hinged sliding cover to protect against moisture. It is usually advisable to install a hopper drier, especially when processing certain materials such as hygroscopic plastics, regrind, and colors. It can be of value in limiting material handling, as well as in removing moisture. Different equipment designs are used for dispensing, metering, and mixing, which use volume and the more popular, useful, and cost-saving gravimeter/weight blenders located over the feed hopper. Motor-driven augers as well as air-driven valves can be included to process materials such as flakes, powders, granular material, liquids, and pellets.

Plastic usage for a given process should be measured so as to determine how much plastic should be loaded into the hopper. The hopper should hold enough plastic for possibly 30 minutes to 1 hour of production. This action is taken so as to prevent storage in the hopper for any length of time.

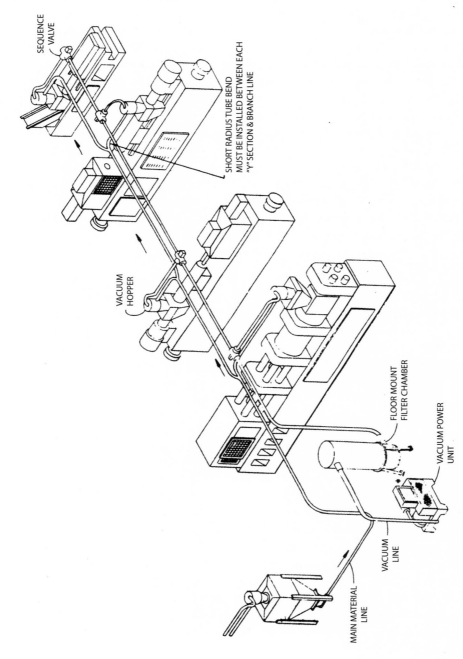

Figure 18.17 Example of a single pneumatic material-handling line-feeding hoppers.

Feeding System Selection Work Sheet

Material: _____   Friable: yes _____ no _____

Bulk density: _____ lb/ft³ (kg/m³)   System voltage: _____

Process throughput: maximum _____ lb (kg)/h   minimum _____ lb (kg)/h

1. Additive throughput: maximum _____ lb (kg)/h   minimum _____ lb (kg)/h
2. Component throughput: maximum _____ lb (kg)/h   minimum _____ lb (kg)/h
3. Other:

For blending systems, this is the minimal amount of information required for each material component. For single additive systems, if there are plans for the feeder to run more than one type of material, provide this information for each material to be run through the same feeder. Because of the different handling and feeding characteristics of each material, a configuration for one material may not work if the material is changed. Injection molders should also provide the cycle voltage of the injection molding machine (24V or 110V) and the screw recovery time.

Figure 18.17 Example of a single pneumatic material-handling line-feeding hoppers (*continued*).

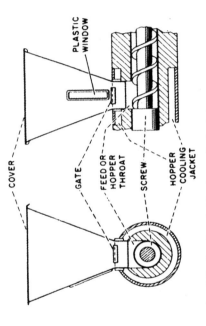

Figure 18.18 Example of the front and side views of a basic hopper.

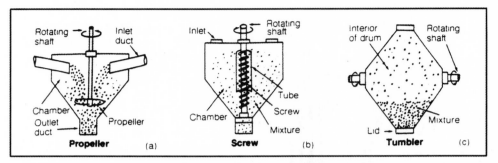

Examples of mixers which receive measured amounts of materials by weight or volume: (a) propeller (impeller) is used in many mixer designs, some suitable for continuous operation; (b) screw (auger) also has many versions. Type shown is carried to the top of a tube and allowed to fall in clouds to the bottom for recirculation until the desired mix/dispersion has been achieved. In other designs the screws are shaped to mix as well as lift the materials so that the tube is omitted; (c) tumblers are batch mixers which receive measured quantities of materials. Type shown is a double-cone mixer/blender.

Figure 18.19 Introduction to hopper mixers.

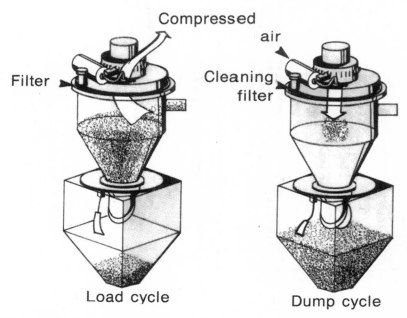

Figure 18.20 Example of a dump-type hopper loader.

Auxiliary and Secondary Equipment 771

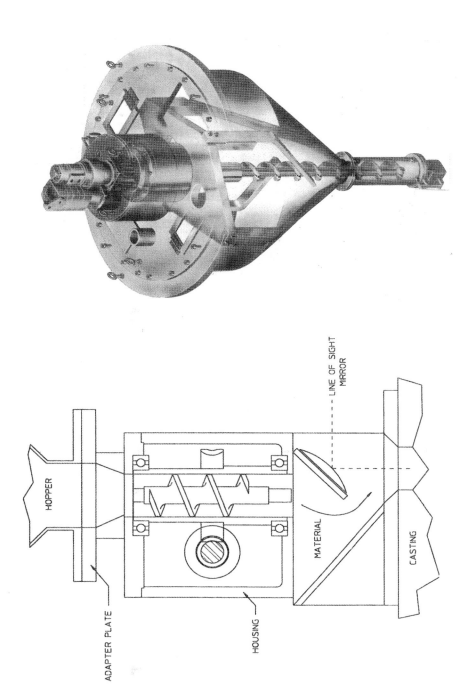

Figure 18.21 Example of a screw-controlled feeding loader (courtesy of Spirex Corporation).

Figure 18.22 Detail view of a hopper screw-controlled feeding loader.

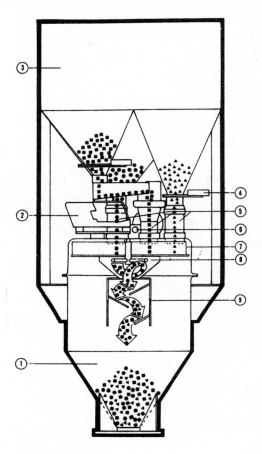

Figure 18.23 Example of components in a hopper blender.

Major Components

1. Machine supply hopper
2. Vibratory additive feeder
3. Material supply hopper
4. Slide gate
5. Material metering tubes
6. Drive motor
7. Metering section
8. Rotating disc
9. Cascade mixing chute

● Virgin
■ Color Concentrate
▲ Regrind

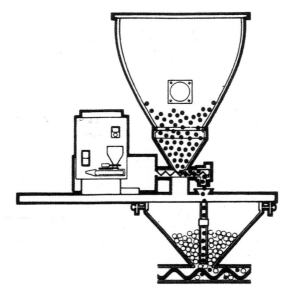

Figure 18.24 Example of metering a color additive in a blender.

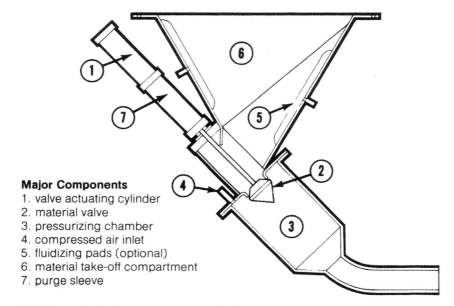

Major Components
1. valve actuating cylinder
2. material valve
3. pressurizing chamber
4. compressed air inlet
5. fluidizing pads (optional)
6. material take-off compartment
7. purge sleeve

Figure 18.25 Example of a hopper power-pump loader.

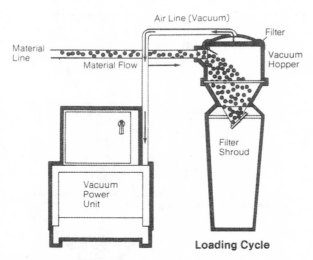

Figure 18.26 Example of a vacuum hopper-loading cycle.

Gravimetric blending improves accuracy and process control and requires less operator involvement (if any) in calibration, particularly when running processes that require great accuracy. Metering by weight eliminates overfeeding expensive additives. A gravimetric feeder or blender has the ability to pinpoint material accuracy despite variations in bulk density. They basically consist of the feeder (including the discharge device), the scale, and the control unit. A separate feeder system is used for blending. There are both batch and continuous units.

Feeding and blending can also be done volumetrically, using belt, rotary, slot, vibrator, or single-screw feeders. Since they adjust by volume, plastics are not self-adjusting for any variations in bulk density. The volumetric has an enclosed chamber that meters a specific volume of materials. Bulk handling and particle size, as well as moisture content, usually influence the uniformity of the metering capacity.

There are vibratory devices for conveying dry materials from storage hoppers to processing equipment, comprising a tray vibrated by mechanical or electrical pulses. The frequency or amplitude or both of the vibrations control the rate of material flow.

Different methods are used to feed plastic through a hopper. They include manual methods to very sophisticated automatic material-handling systems; examples are reviewed in Figures 18.17 to 18.28. Vacuum or positive air-pressure systems are used. The system used depends on factors such as available space, the type of plastic being handled (including its shape and form), blending or mixing requirements, the amount of plastic to be processed, and delivery rate into the plasticator. For example, disc feeders are horizontal, flat, grooved discs installed at the bottom of a hopper feeding a plasticator to control the feed rate by varying the disc's speed of rotation or the clearance between discs. A scraper is used to remove plastic material from the discs. There are stuffers used to handle paste-type molding compounds that do not flow through conventional hoppers. They usually

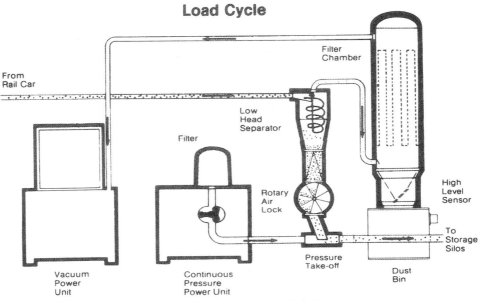
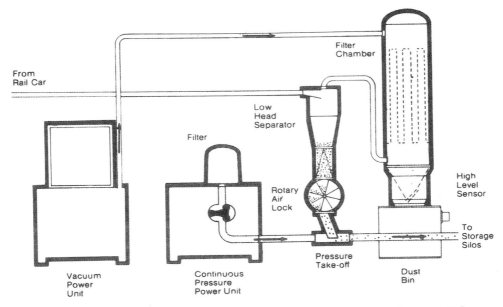

Figure 18.27 Systems utilizing a rotary air lock feeder to separate pressure and vacuum airflow.

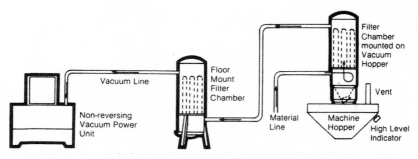

COARSE POWDER FILTER SYSTEM

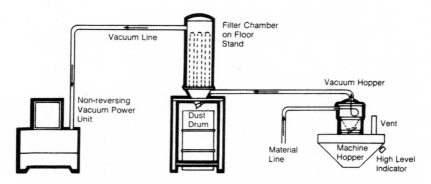

MEDIUM DUSTY MATERIAL FILTER SYSTEM

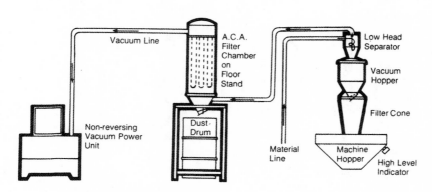

FINE POWDER AND EXTREMELY DUSTY MATERIAL FILTER SYSTEM

Figure 18.28 Examples of coarse, dusty, and powder material-filtering systems.

include a ram or a screw, with the screw also acting as a plunger. Their action is to move material into the plasticator. The stuffer may include a preheater for the material.

There are hopper-mounted coloring feeders and blenders that combine virgin plastics, regrinds, one or more colorants, and additives such as slip agents and inhibitors. Materials are mixed and then the mixture is dropped by gravity into the feed throat. Some coloring loaders allow the use of dry powdered colorants, color concentrates, and liquid colorants through the same unit without major equipment alterations. They are self-loading and mount directly over the processing machines, so there is no need to manually handle component materials and risk contamination and waste. When dry-powder colorants are used, they can be placed in a canister in a separate color room; the filled canister is then mounted on the coloring loader.

Processing TPs is relatively easy compared to TS plastics. Free-flowing TS molding compounds in pellet form, based on plastics such as phenolic, melamine, and urea, can be metered from a hopper just like TP pellets. However, doughy-bulk TS polyester and vinyl-ester compounds, such as bulk molding compound (BMC), require feeding by force. There are basically two ways of feeding these materials to the plastication unit: the batchwise stuffer screw technique and the continuous screw stuffer technique. These types of materials are principally used in injection molding machines.

Hoppers can be fitted with devices to perform different functions. Use is made of hinged or tightly fitted sliding cover for protection against moisture pickup and a magnetic screen for protection against metal ingress. They can be of value in limiting material handling, as well as removing moisture.

Equipment manufacturers have increased the feeding accuracy using different devices such as microprocessor controllers. Materials are being reduced in size with more uniformity to significantly improve uniformity in melt. Processors can use blenders mounted on hoppers that target for precise and even distribution of materials.

Many different machine designs are used to meet the different requirements in the many compounded materials used in the industry. Since no one type of blender (mixer) is likely to satisfy all demands, potential purchasers will have to begin making their buying decisions by determining such factors as degree of type and form of material components, form of material involved, agitation desired, output rate required, and uniformity of output.

INJECTION MOLDING

Various methods are used for handling and moving molded products, depending on factors such as the molding equipment being used, the size and shape of products, setting up for secondary operations, the quantity of products, the system for warehousing, and the system for packaging and shipping to customers. Automating removal of products or parts and other downstream operations reduces processing costs and increases profitability. Automated parts-handling devices can be divided into two categories: take-out devices with transfer mechanisms or gravity systems that receive ejected parts from a mold. There are also robots that perform machine tending and a variety of downstream handling tasks. Figures 18.29 to 18.34 show examples of molded-product take-out with a transfer mechanism or gravity-handling system.

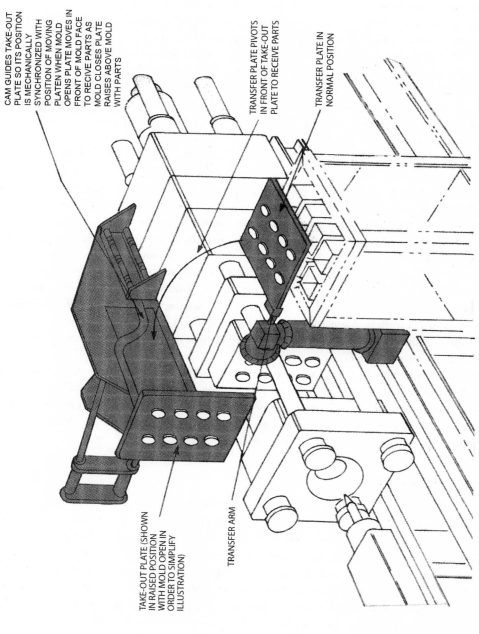

Figure 18.29 Example of a positive take-out and transfer mechanism for molded products (courtesy of Husky Injection Molding Systems Inc.).

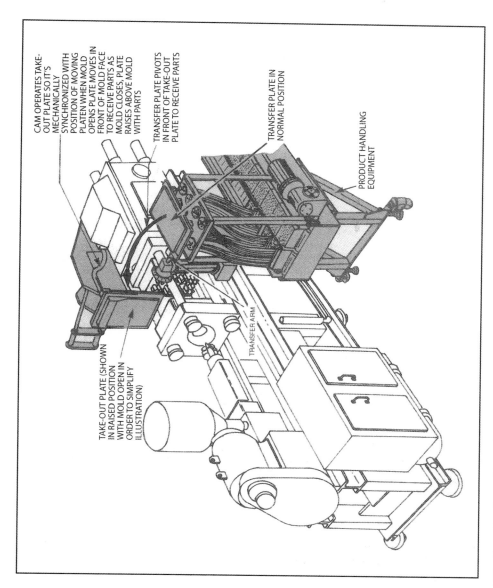

Figure 18.30 Example of a positive take-out system to handle and pack molded products (courtesy of Husky Injection Molding Systems Inc.).

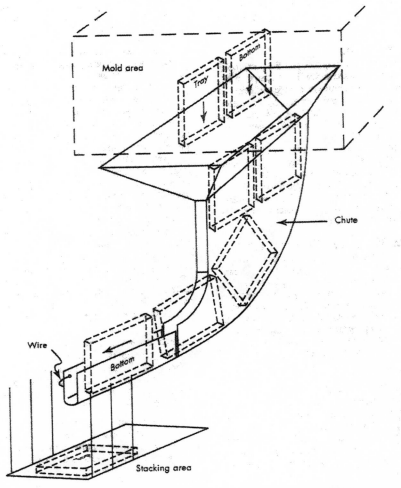

Figure 18.31 Example of a free-drop take-out and transfer mechanism of molded products.

ROBOT

Robots replicate, to various degrees of accuracy, the actions of the human arm and hand. When used for parts removal, they reach into the mold, grasp and remove parts and runners from the mold, and transfer them to the next stage of downstream operations. For simple applications such as machine tending, plastics processors use nonservo robots, in which positioning and speed are controlled mechanically and the sequence of movement is determined by a robot controller. For more complex downstream functions such as sophisticated parts orientation, secondary trimming, hot

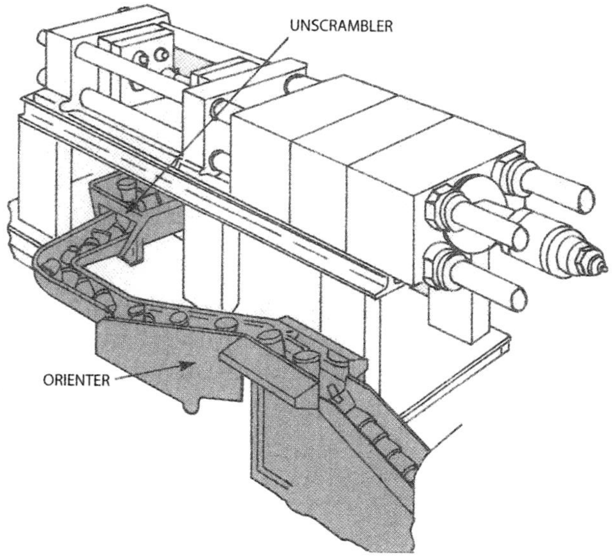

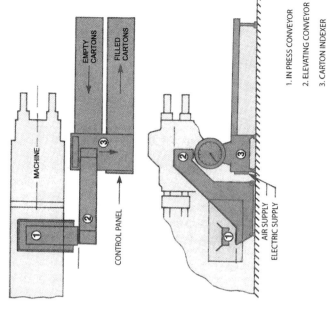

Figure 18.33 Example of bulk filling with automatic carton indexing of molded products (courtesy of Husky Injection Molding Systems Inc.).

Figure 18.32 Example of an unscramble-and-orient system for molded products (courtesy of Husky Injection Molding Systems Inc.).

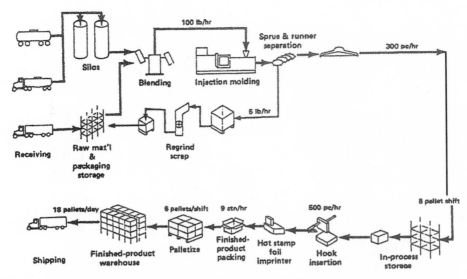

Figure 18.34 Example of flow of material to shipping of molded products.

stamping, and packaging, full-servo robots are used. Position, speed, and sequence are computer-controlled with a feedback closed loop.

A robot arm can have at least six axes of motion, depending on the number of joints in its arm. Normally, the robot provides three perpendicular axes of motion and three additional wrist rotation axes called pitch, roll, and yaw. The combination of all movements controls the positioning of the end-of-arm tooling. Figure 18.35 and Tables 18.6 to 18.8 provide information on robots.

QUICK MOLD CHANGE

Handling raw material quickly and efficiently has been reviewed. This quick and efficient approach also applies to molds, dies, plasticators, and other parts of the production line. To save valuable time—particularly machine downtime—quick changes with microprocessor controls are used in certain plants, replacing manual mold changes (Fig. 18.36). Since most of the quick change occurs in injection molding, the following review concerns quick mold change (QMC; Fig. 18.37).

QUICK MOLD CHANGE DEVICES

QMC devices are used to handle molds (also dies in extrusion lines, dies and molds in blow-molding lines, etc.). The concept is best suited to processors with relatively short production runs and frequent mold changes. The mold change cycle is fully automated for just-in-time (JIT) production. In

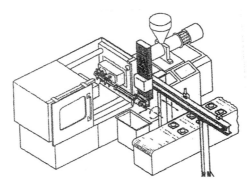

Figure 18.35 Example of a robot removing parts from a mold and depositing them in orderly fashion in a container.

such operations, the benefits of QMC include increased productivity, reduced inventory, increased scheduling flexibility, and more efficient processing. QMC, with microprocessor controls, provides cost-effective approaches to plantwide automation. QMC helps reduce mold-changing time to roughly a couple minutes and facilitates near-instantaneous startup on the new mold.

To make QMC operate with maximum efficiency requires standardizing the construction of molds, machine mounting, and clamping; automation of mold preheat; automatic energy connections via multicoupling plates for centralized connection of nonspill temperature-control and hydraulic circuits and/or electrical power; automatic data connectors; raising operator training levels and improving awareness; and increasing reliance on microprocessor-based controls. Completely automated systems contain the mold conveyors that propel the molds in and out of position on motorized rollers or overhead cranes. The mold carriers are indexed on a track parallel to the molding machine to align the mold conveyors and clamp unit. They can also convey molds to and from different machines or from a central mold storage area.

Different designs include overhead or side-loading and unloading platforms to move molds from an inventory that can have a few to hundreds of molds (Figs. 18.38 and 18.39).

When the microprocessor-based control signals the end of a mold-production run, the movable platen will index to a preset position for mold removal. Automatic mold clamps are deactivated, releasing the old mold. Mold clamps or straps connect to lock together mold halves, and the mold conveyor removes the mold. The computer controls reset platens for the new mold, and the carrier table aligns the new preheated mold with the clamp unit. As mold conveyors insert the new mold, automatic mold positioners center the mold within seconds. Automatic clamps are activated by the control system, and new molding-cycle information stored in the microprocessor initiates the next production cycle. The microprocessor can also make changes in the plastic material being fed to the molding machine. With a highly sophisticated QMC system, estimated downtime between old- and new-mold shot is 2 minutes or less.

Type	Collect	Remove or Pick	Place	Orient	Count/ Weight	Accumulate
Integrated with IMM						
Sweep		×				
Extractor	×	×	×			
Cavity separator	×	×	×			
Robot/bang-bang	×	×	×	×	×	×
Robot/sophisticated	×	×	×	×	×	×

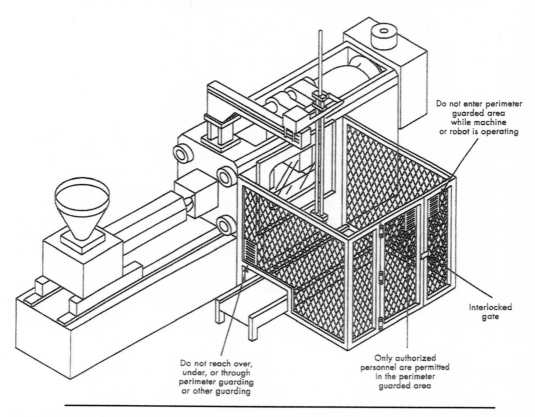

Table 18.6 Examples of the usual functions of robots and perimeter guarding

Types	Features
Pneumatic	For simple, low-cost, point-to-point motion. Typically used for rotary axes. For main robot axes, this drive does not provide sufficient flexibility for advanced automation cells.
Frequency-controlled A/C induction motor	Moderate cost level. Typically used on main robot axes. Used as a lower cost and lower performance alternative to A/C servo. Provides position programming flexibility through the use of a position encoder. Stops and holds position with a mechanical brake. Therefore, long-term reliability is not as good as A/C servo.
A/C servo	Highest cost, fastest payback, lowest overall true cost. Used for both main robot axes and rotary axes. Highest acceleration and speed performance. Greatest positioning accuracy and repeatability. Highest level of reliability. For advanced automation, this is the ideal type of robot drive for the main robot axes.

	Sprue Pickers	All Air	Top-entry Traverse Robots		
			Air with Electric Traverse	3-axis Electric Frequency	3-axis Servo
Features	Pick sprues Simple parts pick	Point-to-point	Multiple stop on traverse	Teach X, Y, Z positions	3-axis
Limitations	Drops high Low payload	Limited stop points	Cannot stack or pelletize	No coordinated motions	None
Benefits	Low cost	Lower cost pick and place	Degate separate placement	More flexibility for downstream	Full flexibility
General market price range	$6,000-9,000	$20,000-24,000	$22,000-28,000	$40,000+	$50,000+
Typical tonnage	30-200 (0.27-1.8 MN)	50-300 (0.4-2.7 MN)	150-300 (1.3-2.7 MN)	2,001 (1.81 MN)	2,001 (1.81 MN)

Table 18.6 Examples of the usual functions of robots and perimeter guarding *(continued)*

	Percentage used with IMM		
Type	Current	Future	Part size
Manual	12	8	Any
Box/collector	20	12	Small, Medium
Conveyor	30	30	Any
Unscramble/orient	10	20	Medium
Sweep	3	5	Small, Medium
Extractor	4	7	Small, Medium
Cavity separator	$\frac{1}{2}$	2	Small
Robot/bang-bang	5	12	Medium
Robot/sophisticated	4	10	Large

Table 18.7 Examples of comparing robots with other parts-handling systems

The computer software stores information regarding platen spacing for each mold change to setting up its process control, which includes molding-cycle times and temperatures. No operator is required to reprogram the plasticator. Increasingly important in software and controls is the ease and efficiency of programming and the ability to change programs quickly, with more microprocessor storage capabilities, and reliable self-diagnostics.

EXTRUDING

Many different types of AE handle the many extruded products fabricated. These products include sheets, films, pipes, tubings, profiles, wire coatings, and fibers. Figures 18.40 to 18.63 provide examples of the equipment. chapter 5 includes reviews as to where the equipment is used.

Figure 18.40 is a schematic drawing of tension-control equipment for the unwinding substrate. The arrows indicate the motions of the driven tension-control rolls and idlers as well as the substrate and the direction of outward pressure on the rolls.

Figure 18.41 is what is called a "flying splice" on a double-station-unrolling stand. The various parts (a–d) can be described as follows:

a. In the starting position, the substrate is fed into the coater from the old roll A over a bumper roll.
b. The old roll A that will soon be fully unrolled is moved forward, and a new roll B moves onto the stand where the old roll was.
c. Adhesive is applied along roll B near the beginning of the substrate web. The driving rings, located below roll B, are moved against the roll that starts revolving until it has reached the required surface speed.

AUXILIARY AND SECONDARY EQUIPMENT

Table 18.8 Examples of types of robots manufactured

COMPANY	SPRUE PICKERS						TRAVERSE ROBOTS - SIZE						AXIS DRIVE TYPE											
	Load cap. 0-5 lb	Load cap. 5-10 lb	Load cap. 10+ lb	IMM tonn. 0-250 tons	IMM tonn. 250-500 tons	IMM tonn. 500+ tons	Load cap. 0-10 lb	Load cap. 10-100 lb	Load cap. 100+ lb	IMM tonn. 0-1000 tons	IMM tonn. 1000-3000 tons	IMM tonn. 3000+ tons	Vertical, electric	Vertical, pneumatic	Vertical, servo	Horizontal, electric	Horizontal, pneumatic	Horizontal, servo	Traverse, electric	Traverse, pneumatic	Traverse, servo	End-of-arm, electric	End-of-arm, pneumatic	End-of-arm, servo
	1	2	3	4	5	6	7	8	9	10	11	12	13	14	15	16	17	18	19	20	21	22	23	24
ABB Inc., Mfg. & Consumer Ind., New Berlin, WI, (262) 785-3400	■																						■	
AEC/Automation Engineering, Wood Dale, IL, (630) 595-1060	■	■					■	■		■	■									■			■	
Automated Assemblies Corp., Clinton, MA, (978) 368-8914	■	■		■	■		■	■		■	■		■			■			■			■	■	
Battenfeld of America Inc., West Warwick, RI, (401) 823-0700	■	■	■	■	■	■	■	■	■	■	■	■	■		■	■		■	■		■	■	■	
The Conair Group Inc., Pittsburgh, PA, (412) 312-6305	■	■		■	■		■	■		■	■		■		■	■		■	■	■	■	■	■	
Direct-Line Products, Wickliffe, OH, (800) 975-2230	■	■		■	■	■	■	■		■	■		■	■		■	■		■	■		■	■	
Engel, Guelph, ON, (519) 836-0220							■	■		■	■		■		■	■		■	■		■	■	■	
Fanuc Robotics N.A. Inc., Rochester Hills, MI, (800) 477-6268	■	■					■	■	■	■	■	■	■		■	■		■	■		■	■		■
Ferromatik Milacron, Batavia, OH, (513) 536-2428	■	■					■	■		■	■		■		■	■		■	■		■	■	■	
Flow Robotic Systems, Wixon, MI, (248) 669-1101	■																						■	
Geiger Handling USA, Madison, WI, (800) 937-9827			■				■	■		■	■									■			■	
Hekuma GmbH, Etching, Germany, +49 (8165) 633-0							■	■	■	■	■	■	■		■	■		■	■		■	■		■
Husky Injection Molding Systems Ltd., Bolton, ON, (905) 951-5000	■	■					■	■		■	■		■		■	■		■	■		■	■		■
InSol Inc., Simpsonville, KY, (502) 722-1028	■	■					■			■										■			■	
JDV Products Inc., Fair Lawn, NJ, (201) 796-1720	■						■			■										■			■	
Mark 2 Automation/Elwood Corp., Oak Creek, WI, (800) 541-8813	■	■					■	■		■	■		■			■			■			■	■	
Matsui America Inc., Elk Grove Village, IL, (847) 290-9680	■			■	■		■	■		■	■									■			■	
PlastiMatix LLC, Farmington Hills, MI, (248) 478-2100							■	■		■	■		■		■	■		■	■		■	■	■	
Ranger Automation Systems Inc., Shrewsbury, MA, (508) 842-6500	■	■					■	■	■	■	■	■	■		■	■		■	■		■	■	■	
Remak North America, Klöckner Group, Hebron, KY, (606) 525-6610	■	■					■	■		■	■		■		■	■		■	■		■	■	■	
Rixan Assoc. Inc., Dayton, OH, (937) 438-3005								■		■													■	
Sailor USA Inc., Kennesaw, GA, (770) 794-0409	■	■					■	■		■	■		■			■			■			■	■	
Stäubli Corp., Duncan, SC, (864) 433-1980	■	■					■	■	■	■	■	■	■		■	■		■	■		■	■		■
Sterltech Div., Sterling Inc., Milwaukee, WI, (414) 354-0970	■	■					■	■		■	■		■		■	■		■	■		■	■	■	

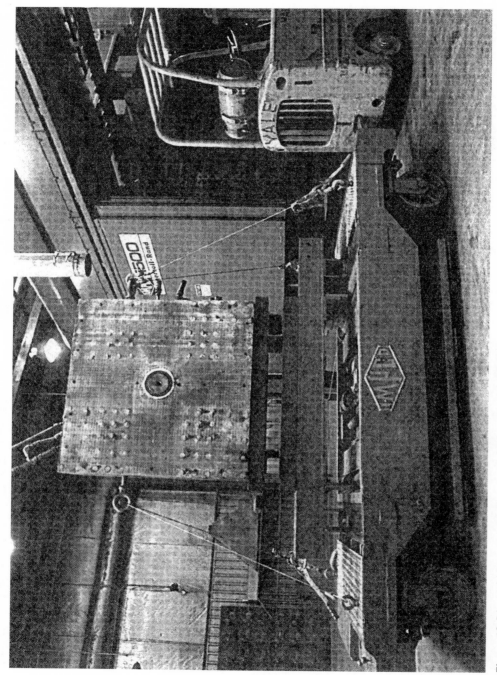

Figure 18.36 Mold base en route manually to injection molding press.

Figure 18.37 Mold base placed manually to the right in injection molding press.

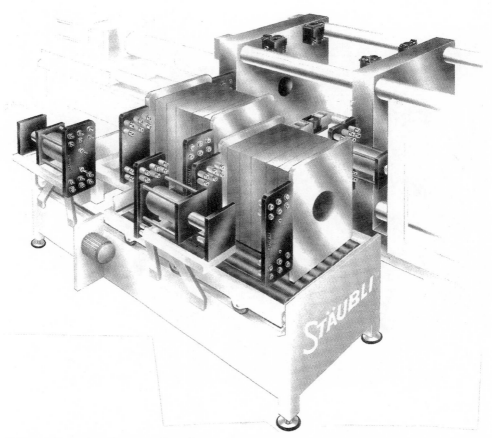

Figure 18.38 Fully automatic horizontal mold change (courtesy of Staubli Corp., Duncan, South Carolina).

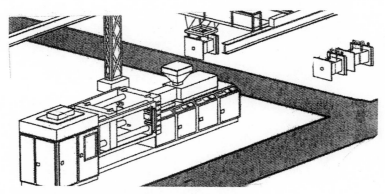

Figure 18.39 Fully automatic overhead-crane mold change.

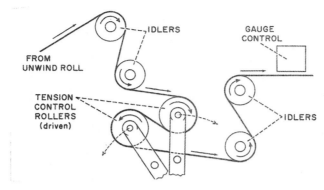

Figure 18.40 Examples of tension-control rollers in a film, sheet, or coating line.

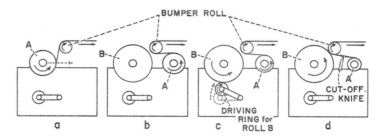

Figure 18.41 Example of laminating with an adhesive.

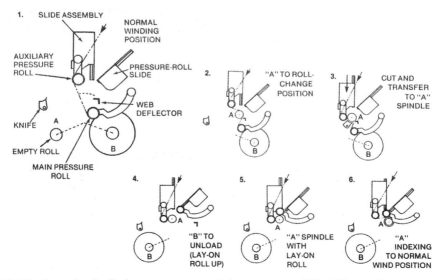

Figure 18.42 Example of roll-change-sequence winder (courtesy of Black Clawson).

Figure 18.43 Closeup view of a tension roll that is processing plastic film.

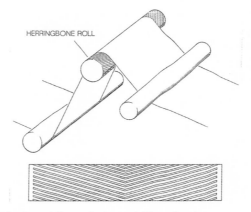

Figure 18.44 Example herringbone idler reducing wrinkles of web.

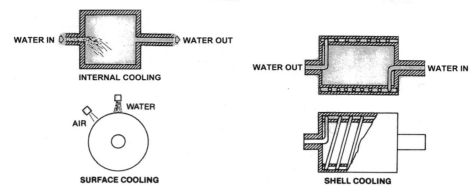

Figure 18.45 Examples of drum-cooling designs with shell cooling being the best design.

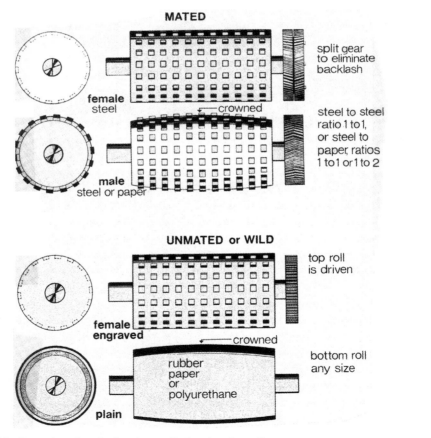

Figure 18.46 Examples of matted and unmatted embossing rolls.

Figure 18.47 Example of a wood-grain embossing roll.

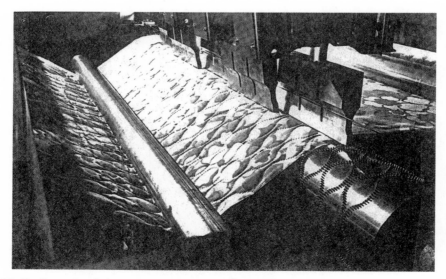

Figure 18.48 Example of ultrasonically sealing a decorative pattern.

AUXILIARY AND SECONDARY EQUIPMENT

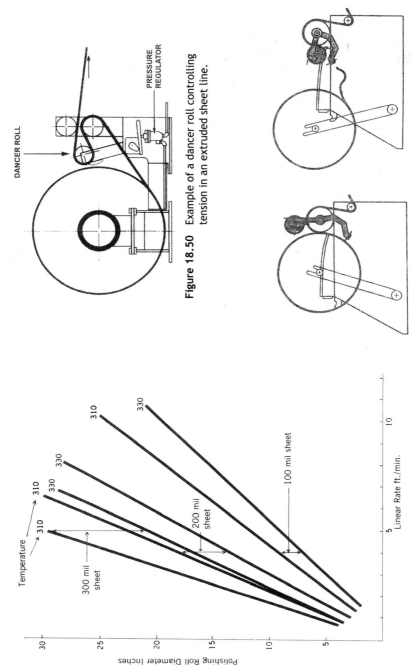

Figure 18.50 Example of a dancer roll controlling tension in an extruded sheet line.

Figure 18.51 Example of an extruded sheet line turret wind-up reel change system.

Figure 18.49 Guide to sheet-polishing roll sizes with a 450°F (230°C) melt temperature.

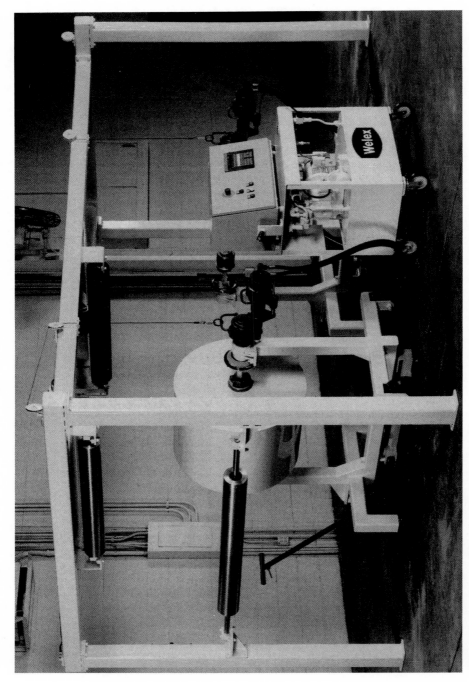

Figure 18.52 View of a large single winder at the end of an extruder sheet line (courtesy of Welex).

Figure 18.53 View of a large dual-turret winder at the end of an extruder sheet line.

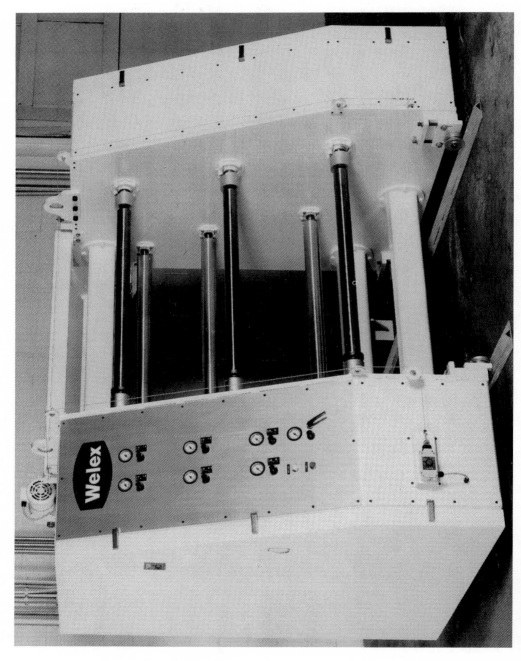

Figure 18.54 View of a sheet roll stock extruder winder with triple fixed shafts (courtesy of Welex).

Figure 18.55 View of downstream extruder-blown film line going through control rolls and dual wind-up turrets (courtesy of Windmoeller & Hoelscher Corporation).

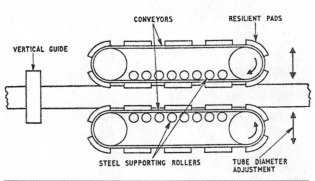

Figure 18.56 Examples of pipe-extrusion caterpillar puller with rollers and conveyor belts.

d. Roll B is moved forward until it contacts the bumper roll. Since now both rolls A and B rotate, substrates from both rolls are bonded. For a very short time, the doubled substrate layer is fed into the coater. The moment the substrate from roll B has caught on, a cutoff knife immediately moves into position against the substrate. In the meantime, the driving rings are removed from roll B. All the steps detailed in this paragraph must occur almost simultaneously, taking no more than a few seconds.

AUXILIARY AND SECONDARY EQUIPMENT 801

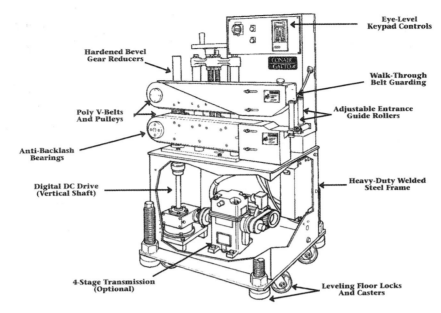

Figure 18.57 Description of a caterpillar belt puller used in an extruder line (courtesy of Conair).

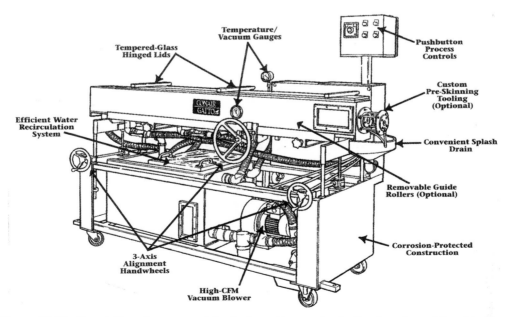

Figure 18.58 Description of a vacuum sizing tank used in an extruder line (courtesy of Conair).

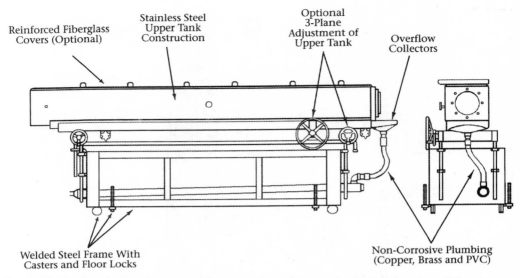

Figure 18.59 Description of a water-and-spray tank used in an extruder line (courtesy of Conair).

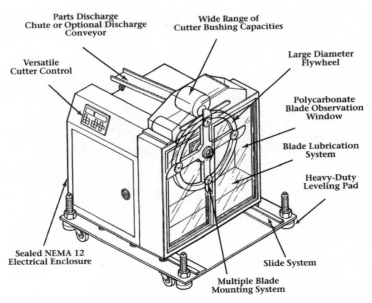

Figure 18.60 Description of a rotary knife cutter used in an extruder line (courtesy of Conair).

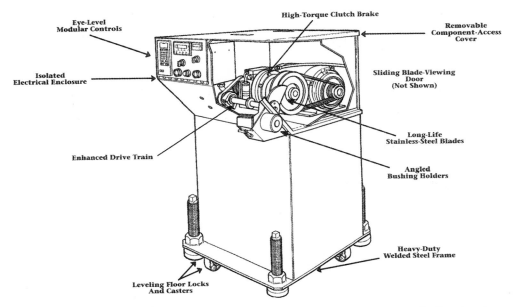

Figure 18.61 Description of a pneumatic-stop rotary knife cutter used in an extruder line (courtesy of Conair).

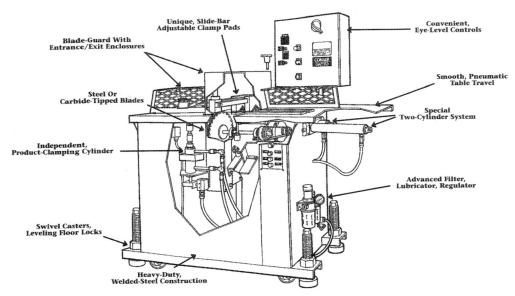

Figure 18.62 Description of a traveling up-cut saw used in an extruder line (courtesy of Conair).

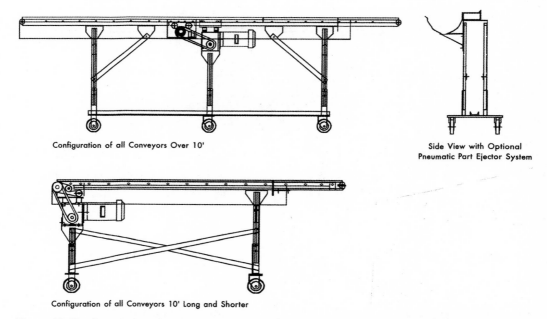

Figure 18.63 Description of a product takeaway conveyor used in an extruder line (courtesy of Conair)

ROLLS

Extruders (and other lines, such as coating lines, which use rolls for windup) require rolls that operate in different ways to perform different functions. Extruder rolls handle films, sheets, coatings, pipes, and other types of lines. They include winder, dancer, lip, spreader, textured, engraved, and cooling rolls. All have the common feature that they are required to be extremely precise in all their measurements, including surface conditions such as commercial-grade mirror finishes, centerlines within tolerance, bearings and all ancillaries mounted on journals, and rotating speed. Control and uniformity of speed is critical in many operations.

Throughputs of winders can be over 2200 lb/h (1000 kg/h). Transfers from one roll to another can take less then a second. Material speeds are up to at least 2200 ft/min (670 m/min) in cast film lines; at least 1000 ft/min (330 m/min) in blown film lines. Blown film lines may want to use reverse-winding systems to allow coextruded films to be wound with a particular material as the inside or outside layer.

Their weights can be very low, going down to at least 16000 lb. Diameters are at least up to 60 in and widths at least up to 30 ft. Some rolls require roundness and surface finishes to be within 0.00005 in (0.00127 mm). Many winders offer sophisticated features and are highly automated, but some are designed to answer the need for simplicity, versatility, and economy. There are surface winders with gap-winding ability for processing tacky films, such as ethylene vinyl alcohols (EVAs) and the metallocene plastics. Information on these different types of rolls is provided in Table 18.9.

DECORATING

An important area in fabricating products is the finishing or decorating of plastics. These steps are usually performed during fabrication or can be performed after fabrication. There are many different methods for adding decorative and functional surface effects, such as printing information on a plastic product (Tables 18.10 to 18.14 and Figs. 18.64 to 18.68). Plastics, of course, are unique in that color and decorative effects can be added prior to and during manufacturing. Pigments and dyes, for example, are compounded into the plastic before the plastic is processed so that color is part of a plastic product and can either be continuous throughout the product or appear on the surface.

Plastic parts can be postfinished in a number of ways. Film and sheet can be postembossed with textures and letterpress or gravure, or silk screening can print them. Rigid plastic parts can be painted or they can be given a metallic surface by such techniques as metallizing, barrel plating, or electroplating. Another popular method is hot stamping, in which heat, pressure, and dwell time are used to transfer color or design from a carrier film to the plastic part. In-mold decorating is also popular; it involves the incorporation of a printed foil into a plastic part during molding so that it becomes an integral part of the piece and is actually inside the piece under the surface. There are applications, such as with blow-molded products, where the foil provides structural integrity, thus reducing the more costly amount of plastic to be used in the products.

Many plastic products are decorated to make them multicolored, to add distinctive logos, or to imitate wood, metal, and other materials. Some plastic products are painted if their as-molded appearance is not satisfactory. This may be the case with reinforced, filled, or foamed plastics.

Common decorative finishing techniques are spray painting, vacuum metallizing, hot stamping, silk screening, metal plating, sputter plating, flame spray or arc spray, electroplating, printing and application of self-adhesive labels, sublimation printing, application of decals, and border stripping. In some cases, the finish will give the product added protection from heat, ultraviolet radiation, chemicals, scratching, or abrasion.

The different decorating methods provide different capabilities and benefits. They include better-quality and second surface graphics, decorations on subtle curves, embossing, design versatility, dry processes that eliminate volatile organic compound (VOC) issues, durability and tamper resistance, economic advantages for multiple colors, elimination of graphics flaws prior to molding, a no-label look, a recessed label area that will not collect much dirt, reduction or elimination of scrap, and elimination of a secondary step for printing.

An important requirement in surface decorations is that the surface be cleaned and prepared in the correct manner for the decorating method used. Some of the common causes of failure in the decorating process are contamination from processing lubricant, dust, natural skin grease, excess plasticizer on the surface, moisture, a frozen strain in the plastic, solvent attack, and so on. The type

Tension Control Roll These type rolls provide the important function of material tension control. There is a proportional relationship between winding tension and lay-on-roll forces (eliminating areas were bumps, valleys, unwanted stretch, etc. develop). There are various tension control techniques available. The proper selection involves decisions on how to produce the tension, how to sense the tension, and how to control the tension. For instance, if the material has a very low tension requirement and if exact control is required, then perhaps, using a magnetic particle brake with an electrical transducer roll with appropriate electronic control is best. However, if the material is on large diameter rolls and moves at slow speed, then a roll follower system can be used effectively.

Dancer Roll These can be used as a tension-sensing device in film, sheet, and coating (wire, film, etc.) lines. They provide an even controlled rate of material movement. Type roll can have an influence on the roll's performance. As an example, chrome plated steel casting drums would seem to be very durable dancers. If used in the absence of a nip roll, should last many years. However, these rolls are in fact very soft due to the annealing which good rolls receive for stress relieving the steel.

As an example a casting drum can been coupled with a steel chill roll to nip polish a cast film web. The casting drum was imprinted by hard plastic edges or die drips. This action occurs because the compressive stresses in a solid plastic passing through the nip of the rolls will exceed the yield strength of the relatively soft steel drum surface. Higher line speeds make the problem worse. In order to prevent this damage, the roll must be hardened.

Adjustable Roll The dancer rolls, canvas drag brakes, various pony brakes, and pneumatically operated brakes are manual adjustment systems. The most expensive would be the regenerative drive systems. The transducer rolls and dancer rolls would be a close second. These systems are usually required in high web speed applications where accurate tension control of expensive and/or sensitive material is paramount. With roll windup systems different roll or reel-change systems are used to keep the lines running at their constant high speeds.

Decorating Roll When the melt leaves the die and enters roll nips, it is soft enough to take the finish of the rolls it contacts. Thus, in addition to smooth and highly glossy finishes, textured or grain rolls can be used. They can impart a mirror image. They can give both functional and aesthetic qualities to the film or sheet. There are as many different grains as the imagination can conjure up.

Cooling Roll These systems range from very inexpensive with rather poor surface non-uniform temperature control to the usually more desirable (and expensive) rolls suitably cored to permit controlled circulation of cool water providing very uniform surface temperature control. When sizing chillers, be sure to include the heat generated by the pumps. For example, a 20 hp water pump can require up to 4 tons of additional chilling capacity to remove the heat generated by the pump.

Spreader/Expander Roll These are bowed rolls that stretch film to remove wrinkles. They create an ever-increasing skew or angle on the roller's rotation, providing a shift in web direction from the roll's center outwards toward the ends. Its major benefit is that the roller "crown" or skew can be adjusted while the line is running to shift the orientation of the web as it passes over the roller. However, the bowed roller design can alter the natural flow of the web, creating uneven tension across the face of the roller, resulting in possible drag in the processing line. This action can cause the web to stretch and distort, especially with thin films. They require a specific amount of space to provide optimum performance.

These grooved rolls have opposing, etched spiral grooves that start at the roll's center and spiral toward the ends. As the roll turns, air flows and follows the direction of the grooves along the metal surface moving from the center of the roll outward. This action forces any web wrinkles out towards the ends of the roll. The expander film spreader roller can consist of a flexible center shaft, a series of bearings placed along the shaft, a flexible metal inner covering, and a smooth-surfaced, one-piece elastomer outer covering.

There is the stretchable one-piece rubber sleeve supported by a series of brushes. As the roll rotates, the entire roller sleeve, as opposed to individual cords, expands and contracts to provide spreading action. The two factors of the wrap or angle at which the web enters onto the roller and the angular displacement of the end caps control the amount of spreading. Notable advancements in this expandable sleeve roller include a smooth, continuous surface that does not produce marking or allow air to enter under the web. Unfortunately the stretching of the rubber can cause the roller to eventually wear over time.

Winding Strain Roll Winding strain can occur at the end of the line. It is the phenomena of a wound roll of film turning into hard rock corrugated nightmare in a few days. This action is caused by several factors: (1) Trapped air as the roll is being wound makes a roll feel soft. Static charges helps trap air. Lay-on roll help to

Table 18.9 Examples of different rolls used in different extrusion processes

squeeze air out but can also create other problems. The rapid escape of air can produce telescoping. (2) Tension creates a compression load, which will squeeze out the very thin film of air, crush under-layers, and crush cores. Tension also tends to even out some of the wrinkles and irregularities. (3) Room temperature recoverable strains are residual processing strains that will release themselves at room temperature to produce a stress and/or shrinkage. Available are techniques for predicting the level of room temperature recoverable strains. (4) Crystallization of crystalline plastics also produces shrinkage of a magnitude generally ½ to 2%. Crystals take less space and thus, as the crystal structure goes to completion, shrinkage occurs. It is permitted to shrink for about one to two days, slit, and rewound.

Preheater Roll As an example in a coating line a heated roll is installed between a pressure roll and unwind roll. Purpose is to heat the substrate prior to being coated.

Table 18.9 Examples of different rolls used in different extrusion processes *(continued)*

of release agent used during the fabrication of products can cause poor bonding of decorations or affect the finish on the product's surface. Zinc stearates are the least harmful with silicones that could be very damaging. Frequently, it is necessary to preheat or pretreat a plastic surface (using flame or chemicals) before applying the decoration. Static electricity on surfaces tends to cause major problems; it attracts airborne impurities, which settle on the surfaces. There is equipment that will avoid static charges. Decoration of any material should be carried out in a clean room under controlled conditions.

JOINING AND ASSEMBLING

Plastic parts can be joined or assembled to other plastic parts of similar or dissimilar plastic materials as well as other materials such as metals. It is often necessary when (1) the finished assembly is too complex or large to fabricate in one piece, (2) disassembly and reassembly is necessary for cost reduction, or (3) different materials must be used in the finished assembly. The ideal situation continues to be in joining or assembling during fabrication whenever it can be done; this helsp to significantly reduce time and the cost of the composite product. When joining and assembling during secondary operations and during fabrication, different factors. such as the coefficient of expansions of the different materials, have to be considered (chapter 19).

Different processes are used for joining and assembling parts; they include using adhesives, using solvents, mechanical tactics, and welding. Tables 18.15 to 18.17 provide an introduction to these methods. Table 18.18 provides a directory of companies that provide these joining and assembling services.

Adhesive and Solvent Bonding

In adhesive and solvent bonding, a material different from either of the parts to be joined is applied to the joint surfaces. In solvent bonding, a solvent is used to dissolve the joint surfaces of plastic parts; the parts are then held together as the solvent evaporates, forming a bond. In adhesive bonding, an adhesive applied to a joint develops structural features as it cures, forming a bond to both joint surfaces. Adhesive bonding is the most general of all joining techniques and can be used to join

Process	Method	Equipment	Application	Comment
Painting Conventional spray	Paints sprayed by air or airless gun(s) for functional or decorative coatings. Especially good for large areas, uneven surfaces or relief designs. Masking used to achieve special effects.	Spray guns, spray booths, mask washers often required; conveying and drying apparatus needed for high production.	Can be used on all materials (some require surface treatment).	Solids, multi-color, overall or partial decoration, special effects such as woodgraining possible.
Electrostatic spray	Charged particles are sprayed on electrically conductive parts; process gives high paint utilization; more expensive than conventional spray.	Spray gun, high-voltage power supply; pumps; dryers. Pretreating station for parts (coated or preheated to make conductive).	All plastics can be decorated. Some work, not much, being done on powder coating of plastics.	Generally for one-color, overall coating.
Wiping	Paint is applied conventionally, then paint is wiped off. Paint is either totally removed, remaining only in recessed areas, or is partially removed for special effects such as woodgraining.	Standard spray-paint setup with a wipe station following. For low production, wipe can be manual. Very high-speed, automated equipment available.	Can be used for most materials. Products range from medical containers to furniture.	One color per pass; multicolor achieved in multistation units.
Roller coating	Raised surfaces can be painted without masking. Special effects like stripes.	Roller applicator, either manual or automatic. special paint feed system required for automatic work. Dryers.	Can be used for most materials.	Generally one-color painting, though multicolor possible with side-by-side rollers.
Screen printing	Ink is applied to part through a finely woven screen. Screen is masked in those areas which won't be painted. Economical means for decorating flat or curved surfaces, especially in relatively short runs.	Screens, fixture, squeegee, conveyorized press setup (for any kind of volume). Dryers. Manual screen printing possible for very low-volume items.	Most materials. Widely used for bottles; also finds big applications in areas like tv and computer dials.	Single or multiple colors (one station per color).
Hot stamping	Involves transferring coating from a flexible foil to the part by pressure and heat. Impression is made by metal or silicone die. Process is dry.	Rotary or reciprocating hot stamp press. Dies. High-speed equipment handles up to 6000 parts/hr.	Most thermoplastics can be printed; some thermosets. Handles flat, concave or convex surfaces, including round or tubular shapes.	Metallics, wood grains or multicolor, depending on foil. Foil can be specially formulated (e.g., chemical resistance).
Heat transfers	Similar to hot stamp but preprinted coating (with a release paper backing) is applied to part by heat and pressure.	Ranges from relatively simple to highly automated with multiple stations for, say, front and back decoration.	Can handle most thermoplastics. A big application area is bottles. Flat, concave or cylindrical surfaces.	Multi-color or single color; metallics (not as good as hot stamp).

Table 18.10 Guide to decorating

Electroplating	Gives a functional metallic finish (matte or shiny) via electrodeposition process.	Preplate etch and rinse tanks; Koroseal-lined tanks for plating steps; preplating and plating chemicals; automated systems available.	Can handle special plating grades of ABS, PP, polysulfone, filled Noryl, filled polyesters, some nylons.	Very durable metallic finishes.
Metallizing Vacuum	Depositing, in a vacuum, a thin layer of vaporized metal (generally aluminum) on a surface prepared by a base coat.	Metallizer, base- and topcoating equipment (spray, dip or flow), metallizing racks.	Most plastics, especially PS, acrylic, phenolics, PC, unplasticized PVC. Decorative finishes (e.g., on toys) or functional (e.g., as a conductive coating).	Metallic finish, generally silver but can be others (e.g., gold, copper).
Cathode sputtering	Uniform metallic coatings by using electrodes.	Discharge systems—to provide close control of metal buildup.	High-temperature materials. Uniform and precise coatings for applications like microminiature circuits.	Metall finish. Silver and copper generally used. Also gold, platinum, palladium.
Spray	Deposition of a metallic finish by chemical reaction of water-based solutions.	Activator, water-clean and applicator guns; spray booths, top- and base-coating equipment if required.	Most plastics. For decorative items.	Metallic (silver and bronze).
Tamp printing	Special process using a soft transfer pad to pick up image from etched plate and tamping it onto a part.	Metal plate, squeegee to remove excess ink, conicalshaped transfer pad, indexing device to move parts into printing area, dryers, depending on type of operation.	All plastics. Specially recommended for odd-shaped or delicate parts (e.g., drinking cups, dolls' eyes).	Single- or multi-color—one printing station per color.
In-the-mold decorating or in-mold labeling	Film or foil inserted in mold is tranferred to molten plastics as it enters the mold. Decoration becomes integral part of product.	Automatic or manual feed system for tie transfers. Static charge may be required to hold foil in mold.	Most plastics, especially polyolefins and melamines. For parts where decoration must withstand extremely high wear.	Single- or multi-color decoration.
Flexography	Printing of a surface directly from a rubber or other synthetic plate.	Manual, semi- or automatic press, dryers.	Most plastics. Used on such areas as coding pipe and extruded profiles.	Single- or multi-color.
Offset printing	Roll-transfer method of decorating. In most cases less expensive than other multicolor printing methods.	Ranges from low-cost hand presses to very expensive automated units. Drying, destaticizers, feeding devices.	Most plastics. Used in applications like coding pipe.	Multi-color print or decoration.
Valley printing	Uses embossing rollers to print in depressed areas of a product.	Embosser with inking attachment or special package system.	Used largely with PVC, PE for such areas as floor tiles, upholstery.	Generally two-color maximum.
Labeling or post-mold labeling	From simple paper labels to multi-color decals and new preprinted plastic sleeve labels.	Equipment runs the gamut from hand dispensers to relatively high-speed machines.	Can be used on all plastics. Used mostly for containers and for price marking.	All sorts of colors and types.

Table 18.10 Guide to decorating *(continued)*

	Economics	Aesthetics	Product Design	Chemistry	Manufacturing	Comments
Appliqué	Unit cost: high	Somewhat limited	Unrestricted	No critical	Hand operation	Allows unusual effects
	Labor cost: high					
	Investment: moderate to high			Good durability		
Electrostatic	Unit cost: low to moderate	Limited	Somewhat restricted	Critical		Dry process, no tool contact with product
	Labor cost: low					
	Investment: moderate to high			Moderate to poor durability		
Flexographic	Unit cost: low	Somewhat limited	Restricted	Critical	Automates well	Wet process, tool contacts product. Sometimes requires top coat
	Labor cost: low					
	Investment: moderate to high		Moderate durability			
Hand painting	Unit cost: high	Somewhat limited	Unrestricted	Critical	Hand operation	Wet process, tool contacts product
	Labor cost: high					
	Investment: low			Good durability		
Heat transfer	Unit cost: low to moderate	Unlimited	Somewhat restricted	Critical	Requires little floor space	Dry process, tool contacts product
	Labor cost: low to moderate					
	Investment: low to moderate			Good durability		Multicolor graphics
Hot stamping	Unit cost: low	Limited	Somewhat restricted	Critical	Requires little floor space	Dry process, tool contacts product
	Labor cost: low to moderate					
	Investment: low to moderate			Good durability		Produces bright metallics
Labeling	Unit cost: low to moderate	Unlimited	Somewhat restricted	Less critical	Adaptable to many situations	Dry process, no tool contact with product at times
	Labor cost: low to moderate					
	Investment: low to high			Moderate to good durability		Multicolor graphics
Metallizing	Unit cost: moderate to high	Limited	Somewhat restricted	Critical	Requires special technological know-how	Wet and dry process, no tool contact with product
	Labor cost: moderate to high					
	Investment: high			Good durability		Produces bright metallics

Table 18.11 Examples of methods for decorating plastic products after fabrication

plastic parts to each other or to dissimilar materials such as metals, aluminum, ceramics, or wood (Tables 18.19 to 18.21). They also can be used to augment the strength of mechanical assembly methods such as snap fits, and can seal assemblies as they are being joined. A range of bond strengths is available (Figs. 18.69 and 18.70). They range from low-strength putty and caulking compounds, which are used only for space and void filling, to high-strength structural adhesives used in many industries (commercial, industrial, aerospace, and aircraft; 5, 309, 310, 402, 579, 580).

Adhesive bonding often is the most efficient, economical, and durable method for plastic assembly. The adhesive can cover the entire bond area and thus can spread stresses rather than concentrate them at the point of fastener attachment. No bosses or holes are required for an assembly

Offset intaglio	Unit cost: low	Limited	Unrestricted	Critical	Requires little floor space	Wet process, tool contacts product
	Labor cost: moderate					
	Investment: moderate			Moderate to good durability		New process
Silk screening	Unit cost: moderate	Somewhat limited	Somewhat restricted	Critical	Flexible operation	Wet process, tool contacts product
	Labor cost: moderate					
	Investment: moderate			Good durability		
Spray	Unit cost: moderate	Limited	Unrestricted	Critical	Requires much floor space	Wet process, no tool contacts with product
	Labor cost: moderate					
	Investment: moderate to high			Good durability		
Woodgraining	Unit cost: high	Specialized	Specialized	Critical	Mostly hand operated	Wet process, tool contacts product
	Labor cost: high					
	Investment: moderate to high			Good durability		
Nameplates	Unit cost: high	Unlimited	Somewhat restricted	Less critical	Adaptable to many situations	Dry process, tool contacts product
	Labor cost: moderate to high					
	Investment: low to moderate			Good durability		Multicolor graphics
Offset	Unit cost: low	Unlimited	Restricted	Critical	Automates well	Wet process, tool contacts product
	Labor cost: moderate					
	Investment: high			Moderate to good durability		Multicolor graphics

Table 18.11 Examples of methods for decorating plastic products after fabrication *(continued)*

	Economics	Aesthetics	Product Design	Chemistry	Manufacturing	Comments
Engraved mold	Unit cost: low	Limited	Unrestricted	Not critical	No extra operations	Best for simple lettering and texture.
	Labor cost: low					
	Investment: moderate			Good durability		
In-mold label	Unit cost: high	Unlimited	Somewhat restricted	Critical	Longer molding cycles	Good for thermoplastics and thermosets. Automatic loading equipment becoming available.
	Labor cost: high					
	Investment: none to moderate			Good durability		
Inserted nameplates	Unit cost: high	Partially limited	Restricted	Not critical	Longer molding cycles	Allows three-dimensional as well as special effects.
	Labor cost: high					
	Investment: moderate			Good durability		
Two-shot molding	Unit cost: high	Limited	Somewhat restricted	Not critical	Two molding operations	Good where maximum abrasion resistance necessary.
	Labor cost: high					
	Investment: moderate to high			Good durability		

Table 18.12 Examples of methods for decorating plastic products in a mold

Decision factors	Hot stamping	Pad printing	Direct screen printing	Heat transfer
Image size and limitations	Roll-on can apply 12 by 24 inches	7 by 14 inches is usual; 10 by 20 opt.	Any size	Roll-on can apply 12 by 24 inches
Resolution of detail	Medium	Fine to medium	Medium	Fine
Arc limits	90° 360° special wrap	100° 360° special wrap	360°	90° 360° special wrap
Opacity	Good	Poor; multiple prints fair	Good	Good for screen; fair for gravure
Wet or dry process	Dry	Wet	Wet	Dry
Part changeover	Seconds to minutes	Minutes to hours	Minutes to hours	Seconds to minutes
Cost of inks, foils, transfers	Foils—cost-sensitive to size; linear increase for addl. colors	Inks—not cost-sensitive to size or color	Inks—not cost-sensitive to size or color	Transfers—cost-sensitive to size; not as sensitive to addl. colors

Table 18.13 Guide in comparing a few decorating methods from size to cost

that is more aesthetically pleasing. This stress diffusion can also lead to the use of thinner and more lightweight sections. Certain considerations must be taken into account to insure that the adhesive bond will be adequate for the final product. Some adhesives may attack or craze plastics, and these adhesives should be eliminated early in testing.

BONDING ACTION

In solvent bonding, the solvent softens the parts being bonded, allowing increased freedom of movement of polymer chains in the plastic (chapter 1). When the two solvent-softened parts are pressed together, molecules from each part come in contact. Van der Waals attractive forces are formed between molecules from each part, and polymer chains from each part intermingle and entangle. As the solvent evaporates, polymer chains become increasingly restricted in movement; after complete evaporation, polymer motions cease, and the amount of entanglement of polymer chains across the bond interface determines bond strength.

In solvent bonding, the interaction between plastic (polymer) and solvent must be maximized. Solvent application must be carefully controlled, since a small difference in the amount of solvent applied to a substrate greatly affects joint strength. Complete evaporation of solvent may not occur for hours or days.

In adhesive bonding, attractive forces form between the adhesive and the adherends (Table 18.22). Types of attractive forces vary with type of adhesive and adherends but are generally a combination of adsorptive, electrostatic, and diffusive forces at the interface between adherend and adhesive. Attractive interfacial forces are very strong; adhesive forces are frequently stronger than

1. Pad Transfer Printing: Wet Process
The decoration of a part by the transfer of ink from a cliche to the part with a silicone rubber transfer pad.

2. Hot Stamping: Dry Process
The application of a single color, metallic, pigment, or pattern foil—such as woodgrain or marble—to a plastic part.

3. Heat Transfers: Dry Process
The application of preprinted, usually multicolor, graphics to plastic parts.
  Note: The hot stamping and heat transfer technologies have the same basic requirements for applications; heat, time, and pressure. Both can be applied by vertical, roll-on or peripheral hot stamp equipment.

4. Silk Screen Printing: Wet Process
Process used to apply inks directly to a part with a squeegee forcing the ink through a fabric or screen.
  General applications are for large, flat areas and cylindrical shaped parts such as blow molded bottles. This process is best for medium to large graphic reproductions. Multi-color applications require a drying/curing station between colors. The advantage of screen printing is in the lay down, or thickness, of the ink. This process is commonly used to produce heat transfers.

5. Gravure or Offset Printing: Wet Process
Wet ink method of printing directly onto the product with very high speeds and a high degree of color-to-color registration.
  Typically, the process yields a thin lay down of ink which is translucent when held to a light. General applications include the food container industry. Multicolor graphics are accomplished by the use of U.V. (Ultraviolet) cure inks. This is also a process used to manufacture heat transfers.
  Note: Pad transfer printing is a form of offset printing.

6. In-Mold Decorating: Dry Process
Offers a finished part directly from the molding press.
  The decorative medium is automatically fed to the molding press or individual inserts, shaped to fit the part, are loaded by hand. The part design should incorporate requirements for "In-Mold" decorating. A high volume of parts is required to justify the costs.

7. Spray Painting: Wet Process
Process for the application of paint or coatings for overall coverage, or select areas by the masking of the part.
  Generally used to improve appearance or part durability. Applications include RFI Shielding for parts in the electronics industry, select area metallizing in the vacuum plating process, and overall coverage on auto body panels.

8. Vacuum Metallizing and Plating: Wet & Dry
Process to apply bright metallic finish to irregular shaped parts.
  Applications include the reflective surfaces of tail and headlight deflectors and the chrome on model parts.

9. Pressure Sensitive Labels: Dry Process
Preprinted, self-adhering paper, foil, or laminated labels.
  Applications include a variety of items in the packaging industry for warning labels, nameplates, instructions, and product identification. Labels can be applied by hand or automatically.

Table 18.14 Review of a few decorating methods

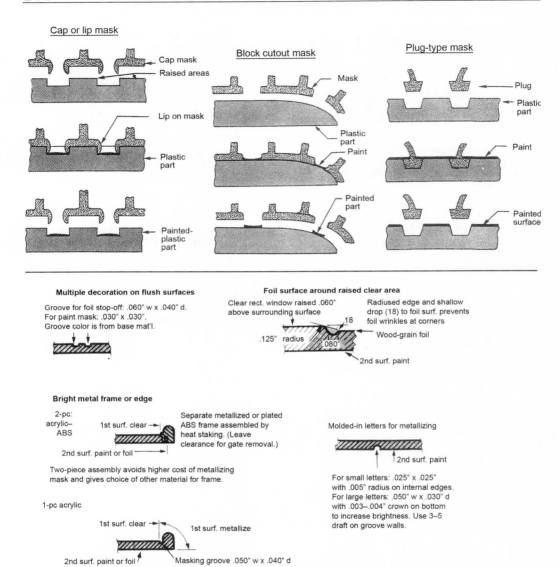

Figure 18.64 Examples in the use of masking for paint spraying.

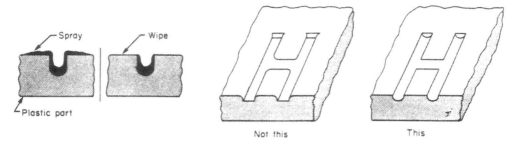

Figure 18.65 Examples of paint spray-and-wipe.

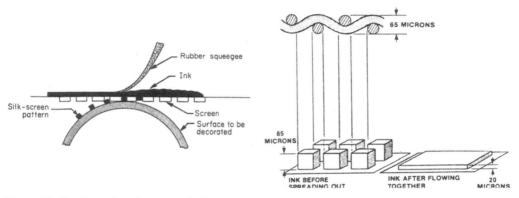

Figure 18.66 Examples of screen printing.

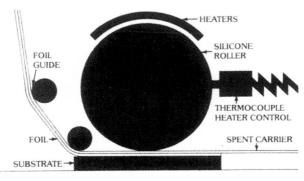

Figure 18.67 Example of hot stamping using a roll-on technique.

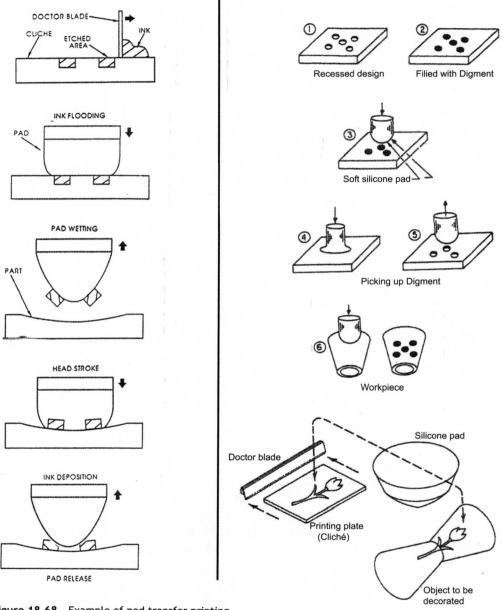

Figure 18.68 Example of pad transfer printing.

	Adhesive bonding	Dielectric welding	Induction bonding	Mechanical fastening	Solvent welding	Spin welding	Staking	Thermal welding	Ultrasonic welding
Thermoplastics									
Polyimide	X			X					
Polypropylenes	X		X	X		X	X	X	X
Propylene copolymers	X		X	X		X		X	
Polystyrenes	X		X	X	X	X	X	X	X
Polysulphone	X			X			X		X
Polyvinyl chloride	X	X	X	X				X	
Polyvinyl chloride copolymers	X	X	X	X				X	
PVC – acrylic compounds	X		X	X					X
PVC – ABS compounds	X	X						X	
Styrene acrylo nitrile	X		X	X		X	X	X	X
Thermoplastic polyesters	X			X	X	X	X		X

Table 18.15 Examples of joining methods

Thermoplastics	Mechanical fasteners	Adhesives	Spin and vibration welding	Thermal welding	Ultrasonic welding	Induction welding	Remarks
ABS	G	G	G	G	G	G	Body-type adhesive recommended
Acetal	E	P	G	G	G	G	Surface treatment for adhesives
Acrylic	G	G	F–G	G	G	G	Body-type adhesive recommended
Nylon	G	P	G	G	G	G	
Polycarbonate	G	G	G	G	G	G	
Polyester TP	G	F	G	G	G	G	
Polyethylene	P	NR	G	G	G–P	G	Surface treatment for adhesives
Polypropylene	P	P	E	G	G–P	G	Surface treatment for adhesives
Polystyrene	F	G	E	G	E–P	G	Impact grades difficult to bond
Polysulfone	G	G	G	E	E	G	
Polyurethane TP	NR	G	NR	NR	NR	G	
PPO modified	G	G	E	G	G	G	
PVC rigid	F	G	F	G	F	G	

Thermosets	Mechanical fasteners	Adhesives	Spin and vibration welding	Thermal welding	Ultrasonic welding	Induction welding	Remarks
Alkyds	G	G	NR	NR	NR	NR	
DAP	G	G	NR	NR	NR	NR	
Epoxies	G	E	NR	NR	NR	NR	
Melamine	F	G	NR	NR	NR	NR	Material notch-sensitive
Phenolics	G	E	NR	NR	NR	NR	
Polyester	G	E	NR	NR	NR	NR	
Polyurethane	G	E	NR	NR	NR	NR	
Silicones	F	G	NR	NR	NR	NR	
Ureas	F	G	NR	NR	NR	NR	Material notch-sensitive

E = excellent; G = good; F = fair; P = poor; NR = not recommended.

Table 18.16 Examples of joining TPs and TSs

Method	Process	Advantage	Limitation
Adhesives			
1. Liquids—Solvent, Water Base, Anaerobics	Solvent- and water-based liquid adhesives, available in a wide number of bases—e.g., polyester, vinyl—in one- or two-part form, fill bonding needs from high-speed laminating to one-piece bonding of dissimilar plastic parts. Solvent types have more bite but cost more. The anaerobics cure in the absence of air and with minimum pressure to effect initial bond.	Ease of application; relatively low cost; high to low bond strengths depending on adhesive and materials being bonded.	Shelf and pot life may be limited; solvents may cause in-plant venting problems. Some may be toxic and must be handled with extreme care.
2. Mastics (e.g., epoxies)	Highly viscous single- or two-component materials which cure to a very hard or flexible joint depending on adhesive type.	Relatively high bond strengths, especially between dissimilar materials. Does not run when applied.	Shelf and pot life often limited. Bond area generally colored.
3. Hot Melts	100% solids that flow when heat is applied. Often used for continuous flat surfaces.	Clean operation, relatively fast.	Virtually no structural strength in plastics applications.
4. Pressure-Sensitive	Tacky adhesives used in a number of applications for a fast bond.	Flexible; often parts can be separated and rejoined.	Bonds not very strong.
Mechanical Fasteners (e.g., staples, screws, molded-in inserts, snaps and a variety of proprietary fasteners)	Often a preferred method of joining dissimilar materials—different plastics or plastics to nonplastics—parts can be specially molded to accept the fasteners. Type selected will depend on strength necessary, appearance factors.	Adaptable to many materials; low-to-medium costs; can be used for parts that may have to be disassembled.	Some have limited pull-out strength; molded-in inserts may result in stresses.
Solvent Cementing & Dopes	Solvent softens the surface of an amorphous thermoplastic; mating takes place when solvent has evaporated. Bodied cement with a small percentage of parent material can give more workable cement and fill in void areas.	Strength up to 100% of parent materials; relatively low equipment costs.	Long solvent evaporation time; solvents may be hazardous; may cause crazing in some resins. Cannot be used for polyolefins and acetal homopolymers.

Table 18.17 Examples of descriptions for different joining methods

Thermal Bonding

1. Ultrasonics	High-frequency sound vibrations generate friction at the bond area of thermoplastic part. Can be used for relatively small-part welds, spot welds or weld and adhesive bond. Most widely used with acetal, ABS, acrylic, nylon, PC, polyimide, PS, SAN, phenoxy.	Strong bonds for most thermoplastics; fast, generally 1-2 sec; good part appearance; clean operation.	Irregular shapes require special handling; only limited use on PVC, polyolefins; no thin parts.	Converter to change 20 KHz electrical energy to 20 KHz mechanical; stand and horn to transmit energy to part; rotary tables or high-speed feeders can be incorporated. Cost for basic equipment is up to $1700 for a portable spot welder; to $5000 for a single-horn standard welder.
2. Hot Plate & Hot Tool Welding	Parts to be bonded are heated against a hot surface until they soften; they are then clamped together at relatively low pressure until bond solidifies. Applicable to rigid plastics.	Very high bond strengths; can be very fast, e.g., 4-10 sec for many jobs.	Stresses may occur in bond area, especially if too high clamp pressure; more complex elements needed for irregular joints.	Ranges from simple soldering guns and hot irons, relatively simple plates up to relatively expensive semi-automatic units.
3. Hot Gas Welding	Welding rod (of same materials being joined) is softened by hot air or nitrogen—which is simultaneously directed through a gun, softening part surface. Rod fills in joint area and cools to effect bond.	Strong bonds, especially for large, structural shapes (it's used most widely with vinyl, but can handle most thermoplastics). Low cost.	Relatively slow (to 60 in./min.); not an appearance weld.	Hand gun, special welding tips, an air source and welding rod. From low-production hand runs to self-contained units. Price range: $85 to $189.
4. Spin Welding	Parts to be bonded are spun at high speed. When bond area reaches softening point, spin stops and parts are held under pressure until bond sets. For most rigid thermoplastics including hi-imp PS, nylon, PE, PVC and the barrier resins.	Very fast—units available for up to 300 containers/min.; strong bonds.	Bond area must be circular; 3:1 length/dia ratio max advisable; need very close part tolerances.	Basic system is spinning device but sophisticated feeding devices are generally incorporated for high-speed operation.
5. Dielectrics	High-frequency voltage applied to film or sheet causes material to melt at bonding surface. Materials cool rapidly to effect bond. Most widely used with vinyls. Also nylon, saran, cellulose acetate.	Fast seal with minimum heat—as fast as 1/4 sec to several secs.	Generally only for film and sheet though some molded parts (on the edges); generally not for polyolefins.	RF generator, dies and press; can range from hand-fed to automatic with automatic indexing. Cost goes from $500 to $100,000 and up.
6. Induction	Metal inserts placed between the parts to be welded and energized with an electromagnetic field. As the insert heats up, the parts around it melt, then cool to form bond. Most thermoplastics.	Rapid heating of solid section to reduce chance of degradation. Fast—2-3 sec max.	Metal remains in plastic so stress may be caused at bond.	High-frequency generator (1-5 KHz used), heating coil and inserts. Hooked to automated feeding and handling devices, processing speeds are high. $2000 to $7500 for power supply. Handling equipment is extra.

Table 18.17 Examples of descriptions for different joining methods *(continued)*

ADHESIVES

3M
Industrial Tape and Specialties Division
3M Center, Building 220-7E-01
St. Paul, MN 55144-1000
(612) 575-5127
(800) 810-1144 (Fax)

trade name(s): VHB Tape

3M
Adhesives, Coatings, and Sealers Division
3M Center
St. Paul, MN 55101
(612) 733-1110

trade name(s): Scotch-weld

Atlas Minerals and Chemicals Corporation
Farmington Road
Mertztown, PA 19539
(610) 682-7171

trade name(s): Atlas

Beacon Chemical Company
125 MacQueston Parkway South
Mount Vernon, NY 10550
(914) 699-3400

trade name(s): Magnacryl

BFGoodrich
Adhesive Systems Division
123 W. Bartges St.
Akron, OH 44311-1081
(216) 374-2900
(216) 374-2860 (Fax)

Cadillac Plastics
1143 Indusco Court
P.O. Box 7035
Troy, MI 48007-7035
(800) 521-4004

trade name(s): PS-30

Ciba-Geigy Corporation
4917 Dawn Avenue
East Lansing, MI 48823
(517) 351-5900
(517) 351-9003 (Fax)

5121 San Fernando Rd. West
Los Angeles, CA 90039
(818) 247-6210
(818) 507-0167 (Fax)

trade name(s): Arathane

Dow Corning Corporation
P.O. Box 0994
Midland, MI 48640-0994
(517) 496-4000
(517) 496-4586 (Fax)

trade name(s): Silastic

Dymax Corporation
51 Greenwoods Road
Torrington, CT 06790
(203) 482-1010
(203) 496-0608 (Fax)

General Electric Corporation
Silicone Products Division
260 Hudson River Road
Waterford, NY 12188
(800) 255-8886

trade name(s): RTV Silicone

H. B. Fuller Company
3530 Lexington Ave. N.
St Paul, MN 55126-8076
(800) 468-6358
(612) 481-1588 (MN)
(612) 481-1828 (Fax)

trade name(s): RA-0018

ITW Devcon
An Illinois Tool Works Company
30 Endicott Street
Danvers, MA 01923
(800) 933-8266
(800) 765-4329 (Fax)
http://www.devcon.com

trade name(s): Superlock, Plastic Welder I & II, PolyStrate, Zip Grip, Flex Welder, Silite, MVP, Plasmetal, Anchorset, 2-Ton Epoxy, 5-Minute Epoxy, Epoxy Plus, 5-Minute Epoxy Gel

Leech Products, Inc.
P.O. Box 2147
Hutchinson, KS 67504-2147
(316) 669-0145

Loctite Corporation
North American Group
1001 Trout Brook Crossing
Rocky Hill, CT 06067-3910
(800) 562-8483
(203) 571-5465 (Fax)

trade name(s): Dri-loc, Fastgasket, Dri-seal, Speedbonder, Shadowcure, Ultra Copper, Ultra Grey, Vibra-seal, Anti-seize, Depend, Impruv, Super Bonder, Resinol, Chipbonder

Lord Corporation
Chemical Products Division
2000 W. Gradnview Blvd.
P.O. Box 10038
Erie, PA 16514
(800) 243-6565
(814) 864-3452 (Fax)

trade name(s): Versilok, Tyrite

Master Bond Inc.
154 Hobart St.
Hackensack, NJ 07601
(201) 343-8983

Table 18.18 Directory of companies that provide joining and assembling methods

National Starch and Chemical Company
10 Finderne Avenue
P.O. Box 6500
Bridgewater, NJ 08807
(908) 685-5000
(908) 685-5096 (Fax)

trade name(s): Bondmaster

Oatey Company
4700-T West 160th
Cleveland, OH 44135
(216) 267-7100

Permabond International
Division, National Starch & Chemical
480 S. Dean Street
Englewood, NJ 07631
(800) 653-6523
(201) 567-3747 (Fax)

trade name(s): Permabond

Permalite Plastics Corporation
1537 Monrovia Avenue
Newport Beach, CA 92663
(714) 548-1137

Norlabs
41 Chestnut St.
Greenwich, CT 06830
(800) 677-5650
(203) 531-0032 (Fax)

Synthetic Surfaces, Inc.
P.O. Box 241
Scotch Plains, NJ 07076-0241
(908) 233-6803
(908) 233-6844 (Fax)

trade name(s): Synthetic Surfaces

Tra-Con, Inc.
P.O. Box 306
Medford, MA 02155
(617) 275-6363

trade name(s): Tra-Con

EXTRUSION WELDING EQUIPMENT

Drader Injectiweld Inc.
8715 50th St.
Edmonton, Alberta T6B 1E7
Canada
(800) 661-4122
(403) 944-0914 (Fax)

HOT GAS WELDING EQUIPMENT

Osram Sylvania
Emissive Products
Portsmouth Ave.
Exeter, NH 03833
(800) 258-8290
(603) 772-1072 (Fax)

Kamweld Products Company Inc.
90 Access Rd.
Norwood, MA 02062
(617) 762-6922

Laramy Products Company Inc.
P. O. Box 1168
Lyndonville, VT 05851
(802) 626-9328
(802) 626-5529 (Fax)

Leister Elektro-Gerätebau
CH-6056 Kägiswil/Switzerland
LEISTER@ACCESS.CH
+41-41-660 0077
+041-41-6602061 (Fax)

Seelye Plastics, Inc.
9700 Newton Avenue South
Minneapolis, MN 55431
(800) 328-2728
(612) 881-6203 (Fax)

HOT PLATE WELDING EQUIPMENT

Branson Ultrasonics Corporation
41 Eagle Road
Danbury, CT 06813-1961
(203) 796-0400
(203) 796-9813 (Fax)

Forward Technology Industries, Inc.
13500 County Road 6
Minneapolis, MN 55441
(612) 559-1785
(612) 559-3929 (Fax)

Sonics & Materials Inc.
Kenosia Avenue
Danbury, CT 06810 (USA)
(800) 745-1105
(203) 798-8350 (Fax)

22 ch du Vernay, Case Postal 627
CH 1196 Gland, Switzerland
(European Office)
41 22-364 1520
41 22-364 2161 (Fax)

sonicsmatl@aol.com

Ultra Sonic Seal
368 Turner Industrial Way
Aston, PA 19014
(610) 497-5150
(610) 497-5195 (Fax)

Zed Industries
P. O. Box 458
Vandalia, OH 45377
(513) 667-8407
(513) 667-3340 (Fax)

Table 18.18 Directory of companies that provide joining and assembling methods *(continued)*

INDUCTION WELDING MATERIALS AND EQUIPMENT

Emabond Systems
Specialty Polymers & Adhesives Division
Ashland Chemical Company
(USA)

49 Walnut Street
Norwood, NJ 07648 (USA)
(201) 767-7400
(201) 767-3608 (Fax)

Emabond Europe B.V.
Kruisakkers 26
4613 BV Bergen op Zoom
The Netherlands
01640-55720
01640-41028 (Fax)

trade name(s): Emaweld, Emabond

Hellerbond Division
Alfred F. Leatherman Company
817 Phillipi Rd.
Columbus, OH 43228-1097
(614) 486-7727
(614) 276-9491 (Fax)

INFRARED WELDING EQUIPMENT

Trinetics Group Inc.
530 South Dock St.
Sharon, PA 16146
(412) 346-4140

INSERTION EQUIPMENT (THERMAL INSERTION)

Bryant Assembly Technologies, Inc.
230 Pepe's Farm Rd.
Milford, CT 06460-0948
(203) 878-2929 (CT)
(800) 937-2900
(203) 878-2329 (Fax)

Sonitek
84 Research Drive
Milford, CT 06460
(800) 875-4676
(203) 878-6786 (Fax)

INSERTION EQUIPMENT

Spirol International Corporation
30 Rock Ave.
Danielson, CT 06239
(203) 774-8571

Trinetics Group Inc.
530 South Dock St.
Sharon, PA 16146
(412) 346-4140

INSERTS

Camloc Products Division
Fairchild Fastener Group
3016 W. Lomita Blvd.
Torrance, CA 90505
(310) 784-6650
(310) 784-6646 (Fax)

trade name(s): SpeedSert, RivIT, Camloc CoilThread

Emhart Fastening Teknologies
Automotive Division
P.O. Box 868
Mt Clemens, MI 48046
(810) 949-0440
(810) 949-0443 (Fax)

trade name(s): Dodge, Heli-Coil

Groov-Pin Corporation
1125 Hendricks Causeway
Ridgefield, NJ 07657
(201) 945-6780
(201) 945-8998 (Fax)

trade name(s): Tap-Lok, Form-Lok, Seal-Lok

Heli-Coil
A Black & Decker Company
Shelter Rock Lane
Danbury, CT 06810
(203) 924-4737
(800) 732-3116 (Fax)

Penn Engineering & Manufacturing Corporation (USA)

Old Easton Rd.
P.O. Box 1000
Danboro, PA 18916-1000
(800) DIAL-PEM
(215) 766-3633 (Fax)

PEM International Ltd. (Europe)
Loncaster DN3 1QR
England
01302 886961
01302 885341 (Fax)

trade name(s): IUB, IUTB, IUTC, ISB, ISC, IUC, IBB, IBC, IBLC, ITB, ITC, STKB, STKC, NFPA, NFPA, PPB, PFLB, PKB

Spirol International Corporation
30 Rock Ave.
Danielson, CT 06239
(203) 774-8571
(203) 774-0487 (Fax)

Yardley Products Corporation
10 West College Ave.
P.O. Box 357
Yardley, PA 19067
(215) 493-2723
(215) 493-6796 (Fax)

trade name(s): Trisert, Sharp-sert, Intro-sert, Fiber-sert

LASER SYSTEMS

Convergent Energy, Inc.
1 Picker Rd.
Sturbridge, MA 01566
(508) 347-2681
(508) 347-5134 (Fax)

RADIO-FREQUENCY WELDING EQUIPMENT

Callanan Company
1844 Brummel Dr.
Elk Grove, IL 60007
(800) 732-5123
(708) 364-4373 (Fax)

Table 18.18 Directory of companies that provide joining and assembling methods *(continued)*

High Frequency Technology Company, Inc.
172D Brook Ave.
Deer Park, NY 11729
(516) 242-3020
(516) 242-4823 (Fax)

Kabar Manufacturing Corporation
140 Schmitt Blvd.
Farmingdale, NY 11735
(516) 694-6857
(516) 694-6846 (Fax)

Nemeth Engineering Associates, Inc.
5901 W. Hwy. 22
Crestwood, KY 40014
(502) 241-1502
(502) 241-5907 (Fax)

Sebra Engineering & Research Associates Inc.
Era Plaza
500 N. Tucson Blvd.
Tucson, AZ 85716
(520) 881-6555
(520) 323-9055 (Fax)

Thermex/Thermatron Inc.
60 Spence St.
Bay Shore, NY 11706
(516) 231-7800
(516) 231-5399 (Fax)

RIVETING EQUIPMENT

Bracker Corporation
105 Broadway Avenue
Carnegie, PA 15106
(800) 44-RIVET
(412) 276-4457 (Fax)

Orbitform
1015 Belden Road
Jackson, MI 49204
(517) 787-9447
(517) 787-6609 (Fax)

trade name(s): Orbitform, Adtech

Straub Design Company
2238 Florida Ave. South
Minneapolis, MN 55426
(612) 546-6686
(612) 546-4653 (Fax)

RIVETS

Stimpson Company Inc.
900 Sylvan Ave.
Bayport, NY 11705-1097
(516) 472-2000
(516) 472-2425 (Fax)

SEALING EQUIPMENT

Armac Industries Ltd.
925 Airport Rd.
Fall River, MA 02720
(508) 676-3051
(508) 649-1672 (Fax)

Battenfield Gloucester Engineering Company Inc.
P.O. Box 900
Blackburn Industrial Park
Gloucester, MA 01930
(508) 281-1800
(508) 283-9206 (Fax)

High Frequency Technology Company, Inc.
172D Brook Ave.
Deer Park, NY 11729
(516) 242-3020
(516) 242-4823 (Fax)

Kabar Manufacturing Corporation
140 Schmitt Blvd.
Farmingdale, NY 11735
(516) 694-6857
(516) 694-6846 (Fax)

Sebra Engineering & Research Associates Inc.
Era Plaza
500 N. Tucson Blvd.
Tucson, AZ 85716
(520) 881-6555
(520) 323-9055 (Fax)

Sonitek
84 Research Drive
Milford, CT 06460
(800) 875-4676
(203) 878-6786 (Fax)

Thermex-Thermatron Inc.
60 Spence St.
Bay Shore, NY 11706
(516) 231-7800
(516) 231-5399 (Fax)

Zed Industries
P. O. Box 458
Vandalia, OH 45377
(513) 667-8407
(513) 667-3340 (Fax)

SPIN WELDING EQUIPMENT

Forward Technology Industries, Inc.
13500 County Road 6
Minneapolis, MN 55441
(612) 559-1785
(612) 559-3929 (Fax)

Mastersonics Inc.
12877 Industrial Dr.
Granger, IN 46530
(219) 277-0210

Sonics & Materials Inc.
Kenosia Avenue
Danbury, CT 06810 (USA)
(800) 745-1105
(203) 798-8350 (Fax)

22 ch du Vernay, Case Postal 627
CH 1196 Gland, Switzerland
(European Office)
41 22-364 1520
41 22-364 2161 (Fax)

sonicsmatl@aol.com

Sonitek
84 Research Drive
Milford, CT 06460
(800) 875-4676
(203) 878-6786 (Fax)

Table 18.18 Directory of companies that provide joining and assembling methods *(continued)*

Trinetics Group Inc.
530 South Dock St.
Sharon, PA 16146
(412) 346-4140

Ultra Sonic Seal
368 Turner Industrial Way
Aston, PA 19014
(610) 497-5150
(610) 497-5195 (Fax)

STAKING EQUIPMENT (HEAT STAKING)

Bryant Assembly Technologies, Inc.
230 Pepe's Farm Rd.
Milford, CT 06460-0948
(203) 878-2929 (CT)
(800) 937-2900
(203) 878-2329 (Fax)

Forward Technology Industries, Inc.
13500 County Road 6
Minneapolis, MN 55441
(612) 559-1785
(612) 559-3929 (Fax)

Sonitek
84 Research Drive
Milford, CT 06460
(800) 875-4676
(203) 878-6786 (Fax)

Young Technology
48012 Fremont Blvd.
Fremont, CA 94538
(800) 394-6273
(510) 490-3214 (Fax)

STAKING EQUIPMENT (THERMOSTAKING)

Orbitform
1015 Belden Road
Jackson, MI 49204
(517) 787-9447
(517) 787-6609 (Fax)

trade name(s): ThermuForm

Phasa Developments
Hollands Road Industrial Estate
Haverhill, Suffolk
England CB9 8PU
 (+44) 1440 62014
(International)
01440 714376
(Fax - England)
+44 1440 714376
(Fax - International)

Service Techtonics Inc.
2287 Treat St.
Adrian, MI 49221
(517)-263-0758
(517) 263-4145 (Fax)

T. A. Systems
1873 Rochester Industrial Court
Rochester Hills, MI 48309
(810) 656-5150
(810) 656-8763 (Fax)

SWAGING (THERMAL)

Bryant Assembly Technologies, Inc.
230 Pepe's Farm Rd.
Milford, CT 06460-0948
(203) 878-2929 (CT)
(800) 937-2900
(203) 878-2329 (Fax)

ULTRASONIC EQUIPMENT

American Technology Inc. (AmTech)
25 Controls Drive
Shelton, CT 06484
(800) 888-6089
(203) 877-7685 (Fax)

Branson Ultrasonics Corporation
41 Eagle Road
Danbury, CT 06813-1961
(203) 796-0400
(203) 796-9813 (Fax)

Bryant Assembly Technologies, Inc.
230 Pepe's Farm Rd.
Milford, CT 06460-0948
(203) 878-2929 (CT)
(800) 937-2900
(203) 878-2329 (Fax)

Dukane Corporation
Ultrasonics Division
2900 Dukane Drive
St. Charles, IL 60174
(708) 584-2300
(708) 584-3162 (Fax)

FFR Ultrasonics
Queniborough Industrial Estate
Melton Road
Quenborough Leics LE7 8FP
0533 606016
0533 606017 (Fax)

Forward Technology Industries, Inc.
13500 County Road 6
Minneapolis, MN 55441
(612) 559-1785
(612) 559-3929 (Fax)

Herrmann Ultrasonics

Descostra(e (Industriegebiet)
7516 Karlsbad-Ittersbach
0 72 48/79-0
0 72 48/79 39 (Fax)

630 Estes Avenue
Schaumburg, IL 60193
(USA Office)
(708) 980-7344
(708) 980-1470 (Fax)

Mastersonics Inc.
12877 Industrial Dr.
Granger, IN 46530
(219) 277-0210

Table 18.18 Directory of companies that provide joining and assembling methods *(continued)*

Phasa Developments
Hollands Road Industrial Estate
Haverhill, Suffolk
England CB9 8PU
01440 62014
(England)
(+44) 1440 62014
(International)
01440 714376
(Fax - England)
+44 1440 714376
(Fax - International)

Rinco Ultrasonics AG
Industriestrasse 4
CH-8590 Romanshom-Switzerland
(+41) 071 / 61 16 66
(+41) 071 / 61 16 95 (Fax)

Sonics & Materials Inc.
Kenosia Avenue
Danbury, CT 06810 (USA)
(800) 745-1105
(203) 798-8350 (Fax)

22 ch du Vernay, Case Postal 627
CH 1196 Gland, Switzerland
(European Office)
41 22-364 1520
41 22-364 2161 (Fax)

sonicsmatl@aol.com

Sonitek
84 Research Drive
Milford, CT 06460
(800) 875-4676
(203) 878-6786 (Fax)

Sonobond Ultrasonics
887 South Matlack Street
West Chester, PA 19382
(800) 323-1269
(610) 692-0674 (Fax)

Ultra Sonic Seal
368 Turner Industrial Way
Aston, PA 19014
(610) 497-5150
(610) 497-5195 (Fax)

VIBRATION WELDING EQUIPMENT

Branson Ultrasonics Corporation
41 Eagle Road
Danbury, CT 06813-1961
(203) 796-0400
(203) 796-9813 (Fax)

Bryant Assembly Technologies, Inc.
230 Pepe's Farm Rd.
Milford, CT 06460-0948
(203) 878-2929 (CT)
(800) 937-2900
(203) 878-2329 (Fax)

Forward Technology Industries, Inc.
13500 County Road 6
Minneapolis, MN 55441
(612) 559-1785
(612) 559-3929 (Fax)

Sonics & Materials Inc.
Kenosia Avenue
Danbury, CT 06810 (USA)
(800) 745-1105
(203) 798-8350 (Fax)

22 ch du Vernay, Case Postal 627
CH 1196 Gland, Switzerland
(European Office)
41 22-364 1520
41 22-364 2161 (Fax)

sonicsmatl@aol.com

Ultra Sonic Seal
368 Turner Industrial Way
Aston, PA 19014
(610) 497-5150
(610) 497-5195 (Fax)

Table 18.18 Directory of companies that provide joining and assembling methods *(continued)*

Surfaces	ABS	Acetal	Cellulosic	Ethyl cellulose	Nylon	Poly-carbonate	Polyether-imide	Polyether-sulfone
ABS	4,5,23,25,26	4,16,26	4,16,36	5,16	5,23,25,26	5,25,26,38	5,25,26,38	5,25,26,36
Acetal	4,16,26	4,5,16,23,25,31	4,5,16	4,5,16	5,23,25,26	4,16,25	4,16,38	5,16,23,25
Cellulosic	4,16,36	4,5,16	4,5,14,16,36	4,5,14,16,36	5,36	4,16	4,16,38	5,36
Ethyl cellulose	5,16	5,16	5,14,16	4,5,14,15,16,36	5,36	5,16	5,16,38	5,36
Nylon	5,23,25,26	5,23,25,26	5,36	5,36	3,5,22,23,25,26,36	25,26	25,26,38	5,25,26,36
Polycarbonate	5,25,26,38	4,16	4,16	5,16	25,26	4,15,16,25,26,38	3,23,25,38	25,26
Polyetherimide	5,25,26,38	4,16,38	4,16,38	5,16,38	25,26,38	5,23,25,38	5,23,25,38	25,26
Polyethersulfone	5,25,26,36	5,16,23,25	5,36	5,36	5,25,26,36	25,26	25,26	25,26
Polyethylene^c	5,16	5,23,31	5,36	5,15,16,36	5,23,36	15,16	15,16	5,31,36
Polyethylene terephthalate	5,25,26,36	4,5	4,5,36	4,5,36	5,25,26,36	4,25,26	4,25,26	5,13,25,26
Polymethyl methacrylate	25,26	5,31	5,36	5,36	5,25,26,36	25,26	25,26	5,13,25,31
Polyphenylene ether, modified	5,23,25,26,36	4,5,16,23,25	4,36	4,5,36	3,5,23,26,37	4,16,25	4,16,25,38	23,25,36
Polyphenylene sulfide	23,25,26	4,5,16,23,26	4,14,16	4,16	5,22,23,25,26	4,15,16,25	4,15,16,25	23,25
Polypropylene	5,16	5,23,31	5,36	5,36	5,23,36	15,16	15,16	5,31,36
Polystyrene	16,25	5,16,23,25,31	5,16,36	5,16,36	5,23,25,36	16,25,26	16,23,25,26	5,13,25,31
Polyurethane	4,25,36	4,5,23,25	4,5,36	4,5,36	5,23,36	4,25	4,25	5,25,36
Polyvinyl chloride	4,25,26,38	4,5	4,5,36	4,5,36	5,26,36	4,26,38	4,26	5,13,25,26
Tetrafluoro-ethylene	15,23	23	5	5,15	5,22,23	15	15	5
Diallyl phthalate	4,26,31	4,23,31	4,36	4,5,36	3,5,25,26,36	4,25,26	4,25,26	5,25,31,36
Epoxy	4,25,26,38	4,23,31	4,36	4,36	3,23,25,36	4,25,26,38	4,25,26	25,31,36
Melamine	4,16	4,16,23,31	4,16,36	4,16,36	3,23,25,36	4,16	4,16	23,31,36
Nitrile	5,16,25	4,5,16,23,31	4,5,16,36	4,5,16,36	3,5,23,25,36	4,16,25,26	4,16,25,26	5,25,31
Phenolic	23,26,36	4,23,25,31	4,36	4,36	23,25,26,36	4,25,26	4,25,26	25,26,31,36
Urea	4,16	4,16,23,31	4,16	4,16	3,23	4,16,25	4,16,25	25,31

Adhesive number code

Elastomeric
1) Natural rubber
2) Reclaim
3) Neoprene
4) Nitrile
5) Urethane (also thermosetting)
6) Styrene butadiene

Thermoplastic resin
11) Polyvinyl acetate
12) Polyvinyl alcohol
13) Acrylic
14) Cellulose nitrate
15) Polyamide
16) Hot-melt copolymer blends (EVA, EEA, etc.)
25) Cyanoacrylate

Thermosetting resin
21) Phenol formaldehyde (phenolic)
22) Resorcinol phenolic
23) Epoxy
24) Urea formaldehyde
26) Reactive acrylate monomer systems (acrylic)
31) Butyral phenolic
33) Phenolic nylon
36) Polyester (also thermoplastic)
37) Anaerobic
38) Silicone

Miscellaneous
41) Rubber latices (natural or synthetic-water based)
42) Resin emulsions (water-based)

Table 18.19 Examples of adhesives for bonding plastics to plastics

Surfaces	Poly-ethylene	Polyethylene terephthalate	Polymethyl methacrylate	Polyphenylene ether, modified	Polyphenylene sulfide	Polypropylene	Polystyrene	Polyurethane
ABS	5,16	5,25,26,36	25,26	5,23,25,26,36	23,25,26	5,16	16,25	4,25,36
Acetal	5,23,31	4,5	5,31	4,5,16,23,25	4,5,16,23,26	5,23,31	5,16,23,25,31	4,5,23,25
Cellulosic	5,36	4,5,36	5,36	4,36	4,14,16	5,36	5,16,36	4,5,36
Ethyl cellulose	5,15,16,36	4,5,36	5,36	4,5,36	4,16	5,36	5,16,36	4,5,36
Nylon	5,23,36	5,25,26,36	5,25,26,36	3,5,23,26,37	5,22,23,25,26	5,23,36	5,23,25,36	5,23,36
Polycarbonate	15,16	4,25,26	25,26	4,16,25	4,15,16,25	15,16	16,25,26	4,25
Polyetherimide	15,16	4,25,26	25,26	4,16,25,38	4,15,16,25	15,16	16,23,25,26	4,25
Polyethersulfone	5,31,36	5,13,25,26	5,13,25,26,31	23,25,36	25,26	5,31,36	5,13,25,31	5,25,36
Polyethylene[c]	5,15,16,23,31,36,41	5,36	5,31,36	23,25,31,36	5,23	5,23,31,36,41	5,23,31,36	5,23,36
Polyethylene terephthalate	5,36	4,5,13,25,26,36	5,13,25,26,36	4,23,36	5,25,26	5,36	5,13,25,36	4,5,25,36
Polymethyl methacrylate	5,31,36	5,13,25,26,36	2,5,6,13,25,26,31,36	5,25,31,36	5,25,26,36	5,26,31	2,5,6,13,25,31,36	5,36
Polyphenylene ether, modified	23,31,36	4,5,25,36	5,25,31,36	5,13,23,25,26,36	23,25,26	23,31,36	23,25,31,36	4,23,36
Polyphenylene sulfide	5,23	5,25,26	5,25,26,36	23,25,26	23,25,26	5,23,26	5,23,25,36	5,23,25,36
Polypropylene[c]	5,23,31,36,41	5,36	5,26,31	23,31,36	5,23,26	5,15,16,23,31,36,41	5,23,31,36	5,23,36
Polystyrene	5,23,31,36	5,13,25,26	2,5,6,13,25,31,36	23,25,31,36	5,23,25,36	5,23,31,36	2,5,6,13,16,23,25,36	5,23,25,36
Polyurethane	5,23,36	4,5,25,36	5,36	4,23,36	5,23,25,36	5,23,36	5,23,25,36	4,5,23,25,36
Polyvinyl chloride	5,36	4,5,13,26,36	5,13,25,26,36	4,25,36	5,25,26,36	5,36	5,13,25,36	4,5,25,36
Tetrafluoroethylene	5,15,23	5	5	5,23	5,22,23	5,15,23	5,23	5,23
Diallyl phthalate	23,31,36	4,5,25,36	5,25,31,36	3,5,23,31,36	25,31,36	23,31,36	23,25,31,36	4,23,25,36
Epoxy	23,31,36	4,25,36	25,26,31,36	4,23,25,31,37	4,23,25,36	23,31,36	23,25,31,36	4,23,25,36
Melamine	16,23,31,36	4,36	31,36	4,23,31	4,23,36	23,31,36	16,23,31,36	4,23,25,36
Phenolic	5,16,23,31,36	4,5,25,36	5,25,31,36	3,23,25,31,37	4,16,25,26	5,23,31,36	5,16,23,25,31,36	4,5,23,36
Polyester	23,31,36	4,25,26,36	25,26,31,36	4,23,26,36,37	4,25,26,36	23,31,36	23,25,31,36	4,23,25,36
Urea	16,23,31	4,25	25,31	4,23,25,31	3,23,25	23,31	16,23,25,31	4,23,25

Table 18.19 Examples of adhesives for bonding plastics to plastics (*continued*)

Surfaces	Polyvinyl chloride	Tetrafluoro-ethylene	Diallyl phthalate	Epoxy	Melamine	Phenolic	Polyester	Urea
ABS	4,25,26,38	15,23	4,25,26,31	4,25,26,38	4,16	5,16,25	23,26,36	4,16
Acetal	4,5	23	4,23,31	4,23,25,31	4,16,23,31	4,5,16,23,31	4,23,25,31	4,16,23,31
Cellulosic	4,5,36	5	4,36	4,36	4,16,36	4,5,16,36	4,36	4,16
Ethyl cellulose	4,5,36	5,15	4,5,36	4,36	4,16,36	4,5,16,36	4,36	4,16
Nylon	5,23,36	5,22,23	3,5,23,25,36	3,23,25,36	3,23,36	3,5,23,25,36	23,25,26,36	3,23
Polycarbonate	4,26,38	15	4,25,26	4,25,26,38	4,16	4,16,25,26	4,25,26	4,16,25
Polyetherimide	4,26	15	4,25,26	4,25,26	4,16	4,16,25,26	4,25,26	4,16,25
Polyethersulfone	5,13,25,26	5	5,25,31,36	25,31,36	23,31,36	5,25,26,31	25,26,31,36	25,31
Polyethylene[c]	5,36	5,15,23	23,31,36	23,31,36	16,23,31,36	5,16,23,31,36	23,31,36	16,23,31
Polyethylene terephthalate	4,5,13,26,36	5	4,5,25,36	4,25,36	4,36	4,5,25,36	4,25,26,36	4,25
Polymethyl methacrylate	5,13,25,26,36	5	5,31,36	25,26,31,36	31,36	5,25,31,36	23,25,31,36	25,31
Polyphenylene ether, modified	4,25,36	5,23	3,5,23,31,36	4,23,25,31,37	4,23,31	3,23,25,31,37	4,23,26,36,37	4,23,25,31
Polyphenylene sulfide	5,25,26,36	5,22,23	25,31,36	4,23,25,36	4,23,36	4,16,25,26	4,25,26,36	3,23,25
Polypropylene	5,36	5,15,23	23,31,36	23,31,36	23,31,36	5,23,31,36	23,31,36	23,31
Polystyrene	5,13,25,36	5,23	23,25,31,36	23,25,31,36	16,23,31,36	5,16,23,25,31,36	23,25,31,36	16,23,25,31
Polyurethane	4,5,25,36	5,23	4,23,25,26,36	4,23,25,36	4,23,25,36	4,5,23,36	4,23,25	4,23,25
Polyvinyl chloride	4,5,11,13,25,26,36,38,42	5	4,25,36	4,25,36,38	4,36	4,5,36	4,25,26,36	4,25
Tetrafluoroethylene[c]	5	5,15,22,23	23,26	23	23	5,23	23	23
Diallyl phthalate	4,25,26,36	5,23	3,4,5,23,25,26,31,36	3,4,23,25,31,36	3,4,23,31,36	3,4,23,31,36,37	4,23,31,36,37	3,4,23,31,37
Epoxy	4,25,26,36,38	23	3,4,23,25,31,36	3,4,23,26,31,36,38	3,4,23,31,36	3,4,23,31,36	4,23,31,36,37	3,4,23,31,37
Melamine	4,36	23	3,4,23,31,36	3,4,23,31,36	3,4,16,23,31,36	3,4,16,23,31,36	4,23,31,36	3,4,16,23,31
Phenolic	4,5,25,36	5,23	3,4,23,31,36,37	3,4,23,31,36	3,4,16,23,31,36	3,4,5,16,23,25,31,36,37	4,23,31,36,37	3,4,16,23,31,37
Polyester	4,25	23	4,23,31,36,37	4,23,31,36,37	4,23,31,36	4,23,31,36,37	4,23,26,31,36	4,23,31,37
Urea	4,25	23	3,4,23,31,37	3,4,23,31,37	3,4,16,23,31	3,4,16,23,31,	4,23,31,37	3,4,16,23,31,

Table 18.19 Examples of adhesives for bonding plastics to plastics *(continued)*

Thermoplastics	Nonplastics						
	Ceramic	Fabric	Leather	Metal	Paper	Rubber	Wood
ABS	6,16,26, 38	3,5	16,23,26	23,25,26, 38	4,42	5,16,21, 25	4,26,42
Acetal	16,23,26	4,23	4,23,26	4,23	4,16,23	4,16	16,23
Cellulosic	4,16	4,5,42	4,5,42	3,4	16,42	1-5,16	4,16
Ethyl cellulose	14,16	14	14	14	14,16	14,16	14,16
Nylon	4,23,26	3,4	3,4,26	3,23,25, 26	4,41	2,3,25	3,4,26
Polycarbonate	16,23,26, 36,38	23,36	16,23,36	23,25,26, 38	16,36	5,16,25, 36	16,23,26, 36
Polyetherimide	23,38	23,36	16,23,36	23,25,26	16,36	5,16,25, 36	16,23,26, 36
Polyethersulfone	16,23,25	3,4	4,31,36	23,25	5,41	2,6,16, 25	16,26,36
Polyethylene	3,16,41	3,41	3,41	3,31,41	16,41	3,16,41	3,16,41
Polyethylene terephthalate	26,26	5,36	5,26,36	26,36	5,36	13,36	26,36
Polymethyl methacrylate	3,4,26	4	3,4,16, 42	3,4,25	42	1-5	3,4,26, 42
Polyphenylene ether, modified	5,24,25	4	16,31,36	23,25,31, 37	4,42	3,4,25	16,26,42
Polyphenylene sulfide	16,23	3,5,42	16,31	4,23,25, 26	4,16,41	5,16,25	16,26
Polypropylene	1,16,41	1,41	1,41	1,2	1,2,16, 41	1,2,16, 41	1,2,16, 41
Polystyrene	16,25,41, 42	3,5	5,31,36	25,31	5,16,31, 36	2,16,16, 25	16,31,36
Polyurethane	4	5,36	4,5	4,5	5,36	5,36	36
Polyvinyl chloride	4,5,26, 38	4,5,42	4,5,16,41, 42	3,4,15,26, 36,38	42	4,5,15	4,26,36, 42
Tetrafluoroethylene	23	22	22	22, 23	22,23	23	23

Adhesive number code is shown in Table 18.19

Table 18.20 Examples of bonding TPs to nonplastics

Thermosets	Nonplastics						
	Ceramic	Fabric	Leather	Metal	Paper	Rubber	Wood
Diallyl phthalate	5,24,37	36	26,31,36	25,31,37	31,36	25,31	26,31,36
Epoxy	23,31,37, 38	4	4,26	23,25,31, 37,38	4	4,25	23,31
Melamine	3,16,26	4	3,4,26	4,26	16,41,42	2,3,4,16, 25	3,16,26
Phenolic	3,16,37	4	3,4,26	3,25,37	16,42	2,3,4,16, 25	3,16,26, 42
Polyester	3,26	4	5,26	5,26,37	41	1-5,25	3,26
Urea	4,16	4,42	3,4,26	3,4,37	16,42	1-5,16,25	3,16,42

Adhesive number code is shown in Table 18.19

Table 18.21 Examples of bonding TS plastics to nonplastics

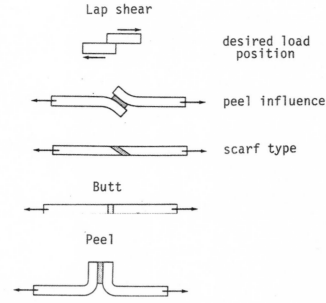

Figure 18.69 Joining and bonding methods.

the cohesive forces of the adhesive or the adherend, and joint failure generally occurs in the adhesive or adherend, not at the joint interface.

Cure time and method are important in obtaining optimum joint strength. Although the classical definition of curing a plastic refers to the chemical cross-linking of a TS polymer (chapter 1), a broader definition is used for the curing of adhesives. Adhesive curing generally refers to the process in which a plastic undergoes a change from the liquid to the solid (gel, rubber, or hard plastic) rheological state, regardless of the physical or chemical method used to induce the change. The method used to induce the liquid-to-solid conversion depends on the class of adhesive. For a TP adhesive, curing is the physiochemical process of cooling the molten plastic. Other curing processes involve chemical reactions; TS adhesives (epoxies, polyurethanes [PUR], and others) are cured through chemical cross-linking, while other adhesives, such as cyanoacrylates, cure by polymerization of a liquid monomer.

Some adhesives cure by exposure to high temperatures; however, curing agents or initiators are usually necessary to begin polymerization or cross-linking reactions. Many adhesives cure by reaction with weak bases or anionic functional groups (water, amines, anhydrides, and amides). Others require initiators or activators, such as peroxides, oxygen, ultraviolet light, or electron beams, in order to generate free radicals for a polymerization chain reaction. Other adhesives require metal salts, acids, or sulfur for cure reactions. Cures times range from a few seconds to several days.

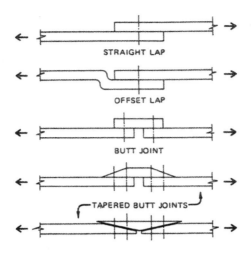

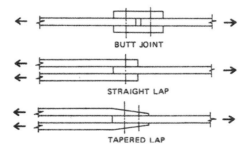

Figure 18.70 Examples of joint geometries.

Adhesive type

A variety of materials are used as adhesives, including natural polymeric compounds (starches, dextrins, proteins, and natural rubber), inorganic materials (silicones), and synthetic polymeric materials (TPs, TSs, and elastomers). Forms of adhesives include pastes, liquids, films, and foams. Adhesives can be classified as anaerobic, which means that they must cure in the absence of air, or aerobic, which means that the curing reactions are not inhibited by oxygen.

Adherend The surface to which an adhesive adheres.

Adherent A body (surface) held to another body (surface), usually by an adhesive or solvent.

Adhesion The state in which two surfaces are held together by the interface forces that may consist of valence forces with interlocking action, or both. The methods or processes used include solvent welding, ultrasonic welding, mixing of reactive components, heat curing, moisture curing, light curing, and surface activation.

Adhesion, mechanical Adhesion between surfaces in which the adhesive holds the parts together by an interlocking action.

Adhesion promoter Also called primer. A coating applied to a substrate prior to adhesive application in order to improve adhesion.

Adhesive, anaerobic Cures only in the absence of air after being confined between assembled parts.

Adhesive assembly The term applied to adhesives used to fabricating finished products as differentiated from adhesives used in the production of sheet materials.

Adhesive bite The ability of an adhesive to penetrate or dissolve the uppermost portions of the adherents.

Adhesive, cold-setting Adhesive capable of hardening at or below room temperature in the presence of a hardener.

Adhesive contact angle The angle formed by a droplet, usually water, in contact with a solid surface, measured from within the droplet. Applicable in determining degree of adhesively bonding or laminating based on wettability of the surface. A contact angle of zero implies complete wetting. A rare occasion is when the angle is 180^0 showing that absolute non-wetting. Intermediate angles correspond to various degrees of incomplete wetting.

Adhesive, cyanoacrylate A highly reactive class of adhesives that cure rapidly at room temperature with a trace of moisture as its catalyst to form high strength bonds with plastics, metals, etc.

Adhesive, heat-active Dry adhesive that is rendered tacky or fluid by the application of heat, or heat and pressure, to the assembly.

Adhesive heat cure A relatively simple process and easily controlled by maintaining consistent cure times and temperature profiles.

Adhesive, hot melt It is applied in a molten state which forms a bond after cooling to a solid state. Acquires adhesive strength through cooling, unlike adhesives that achieve strength through solvent evaporation or chemical cure.

Adhesive, moisture cure These systems polymerize when moisture from the atmosphere diffuses into the appropriate adhesive.

Adhesive, one-part An adhesive that does not require a separate hardener or catalyst for bonding to occur. Types include the use of UV curing, emulsion, solvent, and water/moisture activated.

Adhesive peel strength Adhesive bond strength obtained by a stress applied in a peeling mode.

Adhesive, pressure sensitive An adhesive that requires slight applied pressure on the parts for bonding to occur. Usually composed of a rubbery elastomer and modified tackifier. They are applied to the parts as solvent-based adhesives or hot melts, highly thixotropic, and curing is instantaneous.

Adhesive promoter Also called adhesive primer. A coating applied to a substrate before it is coated with an adhesive to improve the adhesion of the plastic.

Adhesive, room temperature cure Adhesive sets, to handling strength, within at least an hour at temperatures from 20 to 30C (68 to 86F) and latter reaches full strength without heating. A very popular type uses silicone plastic. This room temperature vulcanization (RTV) is the vulcanization or curing at room temperature by chemical reaction, made up of two-part components of silicones and other elastomers/rubbers. RTV adhesives are used to withstand temperatures as high as 290C (550F) and as low as -160C (-250F) without losing their bond strength. Their rapid curing makes them useful in different applications such as adhesives, decorative potting and flexible molds.

Adhesive tackifier A material added to the adhesive to improve the initial and extended tack range of the deposited adhesive film.

Adhesive, two-part An adhesive in which the monomer and catalyst or hardener are separate from each other. The two reactive components separately have an indefinite storage life but must be mixed thoroughly before use.

Table 18.22 Adhesive terminology

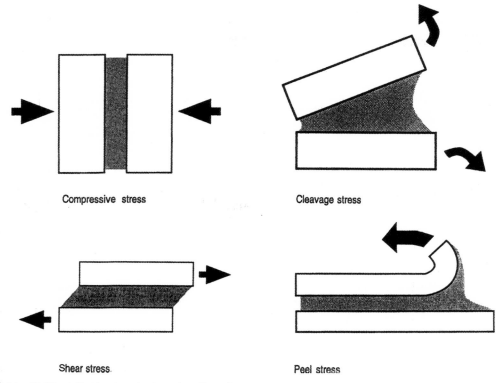

Table 18.22 Adhesive terminology *(continued)*

Functional classifications include holding adhesives, such as masking tape, which hold parts together for limited amounts of time; instant adhesives, which cure within seconds; structural adhesives, which can sustain stresses of at least 50% of the original strength of the part; one-part adhesives, such as rubber cement; and two-part adhesives, such as epoxies.

Adhesives can be water based (meaning that surfactants are used to disperse and stabilize polymer chains into small particles), solvent based, or 100% solids. Use of solvent-based adhesives is declining due to environmental regulations. Most commercial adhesives are classified according to plastic composition. Major classes are described in Table 18.23.

SURFACE PREPARATION

Because it is a chemical-joining process that occurs on the surface, adhesive bonding is very dependent on surface preparation of adherends and the method used for handling, applying, and curing the adhesive. Surface preparation of the joint is necessary for optimum attraction and joint strength.

HOT MELT: Thermoplastics are frequently used as hot melt adhesives. They are applied to the adherends at a temperature much higher than the melting temperature of the thermoplastic, producing a low viscosity fluid that wets the adherend surface. The parts are then clamped, and the thermoplastic adhesive cools and resolidifies, forming a bond between the two substrates. Hot melt adhesives can form both rigid and flexible bonds and can fill gaps and irregularities between parts. Parts must be assembled quickly, before the molten material solidifies. Many hot melts do not possess good wetting properties.

Hot melt adhesives include crystalline thermoplastics (nylons and polyesters, for structural applications), polyethylene for general purpose bonding, polysulfones (high temperature applications), ethylene-vinyl acetate copolymers (low temperature applications), and polyesters. Some thermoplastics and elastomers are applied as one-part emulsions, polymers dispersed in an aqueous solvent, instead of as hot melts; vinyl acetate polymers are commonly used as household white glue. For most applications, one of the substrates must be permeable to allow water removal from the assembly.

ACRYLIC. Acrylic adhesives are derivatives of acrylic acid esters. Long-chain acrylics, such as butyl methacrylate, are relatively soft; methyl methacrylate, with a short carbon chain, has a higher strength and modulus. Blends of short and long chains are also possible, allowing a wide range of flexibility in adhesive properties. Substituted acrylics, such as the cyanoacrylates, are highly reactive and cure within seconds of application. Acrylic adhesives can dissolve grease and wet contaminated surfaces. They are insensitive to adherend surface preparation when compared to other glassy adhesives, such as epoxies.

Curing occurs through polymerization, by free radical or anionic mechanisms. Free radical polymerization, a chain reaction, can be initiated by organic peroxides or other active oxygen compounds or by UV light. Polymerization with peroxide initiators requires elevated temperatures [60- to 100C (140 to 212F)] to occur at a significant rate; activators and catalysts are necessary for polymerization at lower temperatures. Activators, such as reducing agents, induce peroxides to initiate the reaction, while catalysts or accelerators (tertiary amines, polyvalent metal salts) increase the polymerization rate. Maximum operating temperatures are usually 105C (226F), the glass transition of polymethyl methacrylate.

Acrylic adhesives include methacrylates, structural adhesives composed of elastomers and toughening agents dissolved in methyl methacrylate and other monomers, and second-generation acrylics, which contain dissolved polymers to reduce adhesive shrinkage and increase toughness. Cyanoacrylates are highly reactive cyanoacrylic esters that cure rapidly through anionic polymerization at room temperature, with trace amounts of water or amines as catalysts.

Cyanoacrylates can be used for bonding difficult-to-bond plastics such as polyolefins if primers are first applied to the part surface. Primers or adhesive promoters are substances with chemical structures that promote strong adhesive properties (surface tension effects, dipole interactions) when applied in very thin films. They increase adhesion and improve shear, peel, and tensile strength. After evaporation of the primer solvent, the adhesive can be applied; depending on the type of plastic, bond strengths up to twenty times that of the unprimed strength can be obtained. Primers are active from 4 minutes to one hour after solvent evaporation. Cyanoacrylate bonds are frequently stronger than bulk material.

Light-curing acrylics, composed of a blend of acrylic monomers, oligomers, and polymers, polymerize after exposure to ultraviolet light of a particular wavelength and intensity. They are one-part, solvent-free adhesives available in a wide range of viscosities, from 50 cP to thixotropic gels. Cure times generally range from 2 to 60 s. Aerobic adhesives are one-part adhesives composed of dimethacrylate monomers (esters of alkylene glycols and acrylic or methacrylic acid) that cure only in the absence of air. They are less toxic than other acrylics, have a mild, inoffensive odor, and are not corrosive to metals.

EPOXY: Epoxy adhesives are polymers that contain epoxy groups in their molecular structure. Epoxies are usually supplied in the form of liquids or low melting temperature solids, and most contain additives that influence the properties of the material such as accelerators, viscosity modifiers, fillers, pigments, and flexibilizers. Flexibilizers reduce sensitivity to localized regions of high stress by reducing the elastic modulus of the bonded assembly.

Depending on the epoxy, cure can take place at room or elevated temperatures. Two-part epoxies must be mixed just before curing with stoichiometric amounts of amine or anhydride-containing substances. The rate of cure may be increased by adding accelerators to the reaction mixture or by a temperature

Table 18.23 Example of adhesives classified by composition

increase. Heat cured adhesives generally have better mechanical strength than room temperature-cured epoxies. One-part solid epoxies combine epoxy plastics with a curing agent; bonds are rigid and strong, with low peel strength. In epoxy films, a thin film of epoxy and a high temperature curing agent such as an aromatic amine is coated onto release paper. For bonding, the film is cut to the required shape and applied to one of the adherends. Bonds are temperature resistant and brittle and exhibit low peel strength.

ELASTOMER: Natural and synthetic, high molecular weight rubbers or elastomers, dissolved in either hydrocarbon or chlorinated hydrocarbon solvents, are used as adhesives to produce bonds with high peel strength but low shear strength compared to glassy adhesives. They are used when high peel strength is required, such as when bonding a large flexible panel to a rigid composite panel, or as contact adhesives (neoprene or nitrile elastomers), which do not require pressure for bond formation, to bond decorative films onto exterior panels.

Table 18.23 Example of adhesives classified by composition *(continued)*

Joint surfaces should be rough in order to more effectively trap adhesive molecules that bounce across the surface; surfaces should also be free of contaminants that would interfere with the forces of attraction.

Surface preparation cleans the adherend surface or introduces chemical functional groups at the surface to promote wetting and chemical bonding between the adherend and the adhesive. Contaminants include grease, dust, and oil. Generally, surface treatments can be mechanical, chemical, or electrical (Tables 18.24 to 18.26). There are treatments that use LTV light or lasers for surface modification. Cleaning order consists of an initial solvent cleaning, followed by abrasion or a chemical surface alteration. Solvent cleaning is then repeated.

Nonpolar polymers such as polypropylene (PP) are relatively inert, chemically resistant, and difficult to bond. No surface attack takes place at room temperature on molded parts in contact with solvents, so that unless the surface is pretreated, only pressure-sensitive adhesives can be used (Table 18.27). Surface pretreatment increases joint strength considerably. These plastics generally require extensive surface preparation to oxidize the surface and introduce polar reactive groups. Suggested surface preparation methods include etching or oxidation, surface grafting, thermal treatment, using primers, corona discharge (Fig. 18.71), plasma discharge, laser treatment, transcrystalline growth, ultraviolet radiation, flame treatment, or the removal of surface area.

ADHESIVE DATA

Extensive data on the mechanical properties of adhesive bonds per ASTM test methods are available. Tables 18.28 to 18.41 provide data on different plastic materials (579).

MECHANICAL ASSEMBLY

Different types of mechanical fasteners are used in mechanical assembly. They allow simple and versatile joining methods. Mechanical fasteners are made from plastics, metals, or combinations of the two. Fasteners can be removed and replaced or reused when the part on which they are placed

It is a process for modification of plastic surfaces. Results that are more stable and reliable than those of conventional corona discharge processes can usually be achieved with this technique. Additional advantages include the possibility, relatively simply, of adapting it for use with various reactive gases. These create different chemical structures at the plastic surfaces and deliberately generating specific surface properties.

Plasma, a kind of strange state of matter in which gas molecules come apart and become intensely reactive, prepares surfaces prior to bonding, spray printing, and coating plastic parts. Such treatment makes part surfaces harder, rougher, more wettable, less wettable, or easier to adhere. Plasma is also used to treat film and even powder additives like pigments and fillers to make them more dispersible.

Plasma was first used commercially during the 1960s to deposit metal jet coatings on turbine blade for aircraft engines. Plasmas for surface treating plastics go back to the 1970s, but they were used mainly in a handful of high cost value products like contact lenses, electronics, and aerospace components to make them more wettable. Makers of these products kept plasma treatment secret for competitive reasons

Many of the entrepreneurial R&D firms that originally developed plasma-treating applications were brought out in the late 1980s by large companies that were able to bring more financial and engineering clout to developing treating equipment. Equipment is relatively expensive, so it is not for low-value-added commodity products. In some instances, however, it may allow an inexpensive plastic to behave like an expensive one, as an example, making paint adhere to a PP or TPO bumper as well as it would to PUR. For cases were treatment makes economic sense, equipment makers provide an excellent bonding mechanism.

Plasma can be used to treat a range of plastics, including some with notoriously inert surfaces such as polyolefins, engineering thermoplastics, fluoroplastics, and elastomers. Depending on the plastic, it can make bonds 10 to 400 times stronger than can other surface-activation techniques (Tables 18.23 & 18.24). Plasma treatment of plastics has other advantages over the wet-chemical or abrasive processes it can replace, not the least of which is environmental. There are no wet chemicals, fumes, flames, or sand blasting. Instead, plasma spray treating is dry and, depending on residence time, relatively cool from ambient to about 200C (392F) so it does not harm materials that should not get wet or hot.

Table 18.24 Plasma treatment

needs to be serviced. Screws, nuts, and inserts can be made of plastic or metal. Rivets are generally made from metals. In certain molded assemblies, threads are molded in plastic that in turn use screws for joining (Fig. 18.72).

Plastic fasteners are lightweight and resistant to both corrosion and impact. They do not freeze on threads of screws and require no lubrication. Metal fasteners provide high strength and are not affected by exposure to extreme temperatures. Screws, nuts, washers, pins, rivets, and snap-type fasteners are examples of nonintegral attachments, in which the attachment feature is a separate part. Use of separate fasteners requires plastic materials that can withstand the strain of fastener insertion and the resulting high stress near the fasteners (chapters 19 and 20).

Snap-fits are examples of integral attachments, attachment features that are molded directly into the part. Strong plastics that can withstand assembly strain, service load, and possible repeated use are required for nonintegral attachments. Separate fasteners add to production costs due to increased assembly time and requirements of additional material and can be difficult to handle and insert; as a result, use of integral attachment features continues to increase (579).

Table 18.25 Lap shear strength of plastics after plasma treatment per American Society for Testing Materials (ASTM) test methods

Plastic	Bonded to...	Untreated, psi	Plasma treated, psi
Acetal	Self	165	649
ETFE	Self	Low	3200
ETFE	Copper	8	50
FEP	Copper	Low	50
HDPE	Self	315	3500
PPE	Self	617	1800
LCP/25% mineral	Self	940	1240
PC	Self	1705	2242
PC	Aluminum	600	1550
ACM	Self	250	2160
PBT	Self	520	1645
PEI	Self	185	2185
PES	Self	130	3140
PE	Steel	5	22
PI/Graphite	Self	420	2600
PPS	Self	290	1360
PP	Self	370	3080
PS	Aluminum	510	2050

Table 18.26 Peel strength of plastics after plasma treatment per ASTM test methods

Plastic	Untreated lb/in.	Plasma treated, lb/in.
ETFE	0.1	15.8
FEP	0.1	10.5
PI	4	16
PTFE	0.1	2.2
PFA	0.1	8.3
PFA	Low	8.3
RTV Silicone	Low	23
Silicone	0.4	19

Plastic Material Composition (Himont Profax 6323)		BlackMax 380 rubber toughened cyanoacrylate (200 cP)	Loctite Adhesive				
			Prism401 surface insensitive ethyl cyanoacrylate (100 cP)	Prism401 Prism Primer 770 polyolefin primer for cyanoacrylate	Super Bonder 414 general purpose cyanoacrylate (110 cP)	Depend330 two-part no-mix acrylic	Locitite 3105 light cure acrylic (300 cP)
Unfilled resin	5 rms	50 (0.3)	50 (0.3)	>1950 (>13.5)	50 (0.3)	200 (1.4)	100 (0.7)
Roughened	26 rms	50 (0.3)	550 (3.8)	1300 (9.0)	300 (2.1)	200 (1.4)	450 (3.1)
Antioxidant	0.1% Irganox 1010 0.3% Cvanox STDP	50 (0.3)	250 (1.7)	>1950 (>13.5)	50 (0.3)	200 (1.4)	100 (0.7)
UV stabilizer	0.5% Cyasorb UV 531	50 (0.3)	50 (0.3)	>1950 (13.5)	50 (0.3)	200 (1.4)	100 (0.7)
Impact modifier	9% Novalene EPDM	50 (0.3)	150 (1.0)	>1650 (>11.4)	200 (1.4)	200 (1.4)	100 (0.7)
Flame retardant	9% PE-68 4% Antimony oxide	50 (0.3)	50 (0.3)	>1950 (>13.5)	50 (0.3)	200 (1.4)	250 (1.7)
Smoke suppressant	13% Firebrake ZB	50 (0.3)	50 (0.3)	>1950 (>13.5)	50 (0.3)	200 (1.4)	100 (0.7)
Lubricant	0.1% Calcium stearate 24-26	50 (0.3)	50 (0.3)	>1950 (>13.5)	50 (0.3)	200 (1.4)	100 (0.7)
Filler	20% Cimpact 600 Talc	50 (0.3)	50 (0.3)	>1950 (>13.5)	100 (0.7)	200 (1.4)	100 (0.7)
Colorant	0.1% Watchung Red RT-428-D	50 (0.3)	50 (0.3)	>1950 (>13.5)	50 (0.3)	200 (1.4)	100 (0.7)
Antistatic	0.2% Armostat 475	50 (0.3)	200 (1.4)	>1950 (>13.5)	200 (1.4)	200 (1.4)	100 (0.7)

Table 18.27 Shear strength of PP to PP adhesive bonds in psi (MPa) per ASTM D 4501

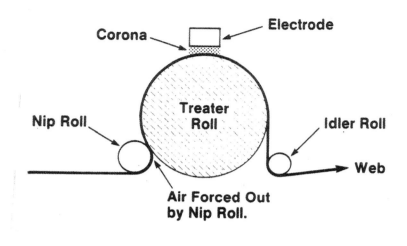

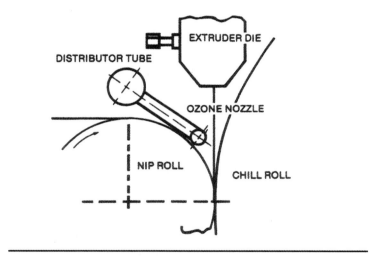

Figure 18.71 Examples of corona treatments in extrusion lines.

Screw assembly

Machine screws, nuts, bolts, and washers are used. They can be disassembled and reassembled an indefinite number of times. Tensile stress should be avoided in assembly design; compressive stress is more desirable due to a lower susceptibility to stress cracking and crazing.

Plastic Material Composition (Dow Chemical 722M LDPE)		Loctite Adhesive					
		Black Max 380 rubber toughened cyanoacrylate (200 cP)	Prism 401 surface insensitive ethyl cyanoacrylate (100 cP)	Prism 401/ Prism Primer 770 polyolefin primer for cyanoacrylate	Super Bonder 414 general purpose cyanoacrylate (110 cP)	Depend 330 two-part no-mix acrylic	Loctite 3105 light cure acrylic (300 cP)
Unfilled resin	5 rms	<50 (<0.3)	150 (1.0)	500 (3.5)	150 (1.0)	150 (1.0)	350 (2.4)
Roughened	88 rms	<50 (<0.3)	150 (1.0)	500 (3.5)	150 (1.0)	150 (1.0)	350 (2.4)
Antioxidant	0.1% Irganox 1010	50 (0.3)	150 (1.0)	500 (3.5)	150 (1.0)	150 (1.0)	350 (2.4)
UV stabilizer	0.3% Cyasorb UV-531	100 (0.7)	100 (0.7)	200 (1.4)	150 (1.0)	200 (1.4)	150 (1.0)
Flame retardant	16% DER-83R 6% Antimony Oxide	100 (0.7)	150 (1.0)	500 (3.5)	150 (1.0)	150 (1.0)	100 (0.7)
Lubricant	1% Synpro 114-36	100 (0.7)	150 (1.0)	500 (3.5)	150 (1.0)	150 (1.0)	350 (2.4)
Filler	17% OmyaCarb F	100 (0.7)	150 (1.0)	500 (3.5)	300 (2.1)	200 (1.4)	350 (2.4)
Colorant	0.1% Watchung Red B RT4280	<50 (<0.3)	100 (0.7)	500 (3.45)	50 (0.3)	150 (1.0)	100 (0.7)
Antistatic	0.4% Armostat 375	<50 (<0.3)	600 (4.1)	500 (3.5)	750 (5.2)	150 (1.0)	200 (1.4)
High density polyethylene	courtesy of Compression Polymers	<50 (<0.3)	50 (0.3)	2000 (13.8)	50 (0.3)	150 (1.0)	100 (0.7)

Table 18.28 Shear strength of polyethylene (PE) to PE in psi (MPa)

Machine screws and bolts should have a flat side under the screwhead (Fig. 18.73). Flathead screws and screws with conical heads produce potentially high tensile stresses due to wedging of the screwhead into the plastic part. Screws with flat undersides, such as the pan-head screw, do not undergo this wedging action, and the stress produced is more compressive. Flat washers distribute the assembly force and should be used under the fastener head.

Screws or threaded bolts with nuts pass through the plastic part and are secured by an external nut or clip on the other side. For compressive rather than tensile loading, space between surfaces of parts being assembled should be eliminated using spacers if necessary. Some part designs may require a loose-fit gap between bosses to prevent high bending stresses or distortion as the parts go into compression.

SELF-TAPPING SCREW

Self-tapping screws are widely used due to their excellent holding force and lower stress in plastics compared to some of the other fastening methods. Self-tapping or self-threading screws are placed in the plastic. No nuts are used, and access is needed from only one side of the assembly. No clear-

Plastic Material Composition (GE Plastics Cycolac GPM 6300)		Loctite Adhesive					
		Black Max 380 rubber toughened cyanoacrylate (200 cP)	Prism 401 surface insensitive ethyl cyanoacrylate (100 cP)	Prism 401/ Prism Primer 770 polyolefin primer for cyanoacrylate	Super Bonder 414 general purpose cyanoacrylate (110 cP)	Depend 330 two-part no-mix acrylic	Loctite 3105 light cure acrylic (300 cP)
Unfilled resin	3 rms	950 (6.6)	>3500[a] (>24.1)[a]	>3350[b] (>23.1)[b]	>3500[a] (>24.1)[a]	300 (2.1)	>3500[b] (>24.1)[b]
Roughened	48 rms	1400 (9.7)	>3500[a] (>24.1)[a]	>3350[b] (>23.1)[b]	>3500[a] (>24.1)[a]	1300 (9.0)	>3500[b] (>24.1)[b]
Antioxidant	0.1% Irgaphos 168 0.16% Irganox 245 0.04% Irganox 1076	950 (6.6)	>3500[a] (>24.1)[a]	>3350[b] (>23.1)[b]	>3500a (>24.1)[a]	150 1.0)	>3500[b] (>24.1)[b]
UV stabilizer	0.4% UV5411 0.4% UV3346 0.1% Ultranox 626	950 (6.6)	>3500[a] (>24.1)[a]	>3350[b] (>23.1)[b]	>3500[a] (>24.1)[a]	300 (2.1)	>3500[b] (>24.1)[b]
Flame retardant	13.5% DE83R 3% Chlorez 700 SS 4% 772VHT Antimony Oxide	950 (6.6)	>3500[a] (>24.1)[a]	>3350[b] (>23.1)[b]	>3500[a] (>24.1)[a]	300 (2.1)	>3500[b] (>24.1)[b]
Smoke suppresant	5% Firebrake ZB zinc borate	650 (4.5)	>3500[a] (>24.1)[a]	>3350[b] (>23.1)[b]	>3500[a] (>24.1)[a]	300 (2.1)	>3500[b] (>24.1)[b]
Lubricant	0.2% N,N'-Ethylene bisstearamide	950 (6.6)	>3500[a] (>24.1)[a]	>3350[b] (>23.1)[b]	>3500[a] (>24.1)[a]	300 (2.1)	>3500[b] (>24.1)[b]
Glass Filler	20% Type 3450 glass fiber	950 (6.6)	>3500[a] (>24.1)[a]	>3350[b] (>23.1)[b]	>3500[a] (>24.1)[a]	300 (2.1)	>3500[b] (>24.1)[b]
Colorant	4% 7526 colorant	950 (6.6)	>3500[a] (>24.1)[a]	>3350[b] (>23.1)[b]	>3500[a] (>24.1)[a]	300 (2.1)	>3500[b] (>24.1)[b]
Antistatic	3% Armostat 550	>3500[a] (>24.1)[a]	>3500[a] (>24.1)[a]	>3350[b] (>23.1)[b]	>3500[a] (>24.1)[a]	300 (2.1)	>3500[b] (>24.1)[b]

[a] The force applied to the test specimens exceeded the strength of the material resulting in substrate failure before the actual bond strength achieved by the adhesive could be determined.
[b] Due to the severe deformation of the block shear specimens, testing was stopped before the acutal bond strength achieved by the adhesive could be determined (the adhesive bond never failed).
[c] All testing was done according to the block shear method (ASTM D4501).

Table 18.29 Shear strength of ABS to ABS in psi (MPa)

ance is required, since the mating thread fits the screw threads. Only a drilled or molded pilot hole is required, and assembly is rapid.

Thread-forming screws form threads in the plastic by displacing and deforming plastic material, which flows around the screwheads. No material is removed, which creates a fit with zero clearance. Thread-forming screws are inexpensive and vibration resistant. They are recommended for solid parts and structural foam parts. Although they can provide acceptable performance in solid-wall parts, they should be used with caution in solid materials due to the possible generation of highly stressed regions. Multiple assemblies and disassemblies are possible with thread-forming screws.

Thread-cutting screws use preformed hole depths that are slightly longer than the screw depth in order to provide a depository for the plastic chips. High internal stresses are avoided in thread-cutting screws due to the removal of material during installation. Stress relaxation of the plastic with temperature changes and time has only a minimal effect on screw performance. Only a minimum

Plastic Material Composition (Himont Profax 6323)		Loctite Adhesive					
		Black Max 380 rubber toughened cyanoacrylate (200 cP)	Prism 401 surface insensitive ethyl cyanoacrylate (100 cP)	Prism 401/ Prism Primer 770 polyolefin primer for cyanoacrylate	Super Bonder 414 general purpose cyanoacrylate (110 cP)	Depend 330 two-part no-mix acrylic	Loctite 3105 light cure acrylic (300 cP)
Unfilled resin	5 ms	50 (0.3)	50 (0.3)	>1950[b] (>13.5)[b]	50 (0.3)	200 (1.4)	100 (0.7)
Roughened	26 ms	50 (0.3)	550 (3.8)	1300 (9.0)	300 (2.1)	200 (1.4)	450 (3.1)
Antioxidant	0.1% Irganox 1010 0.3% Cyanox STDP	50 (0.3)	250 (1.7)	>1950[b] (>13.5)[b]	50 (0.3)	200 (1.4)	100 (0.7)
UV stabilizer	0.5% Cyasorb UV 531	50 (0.3)	50 (0.3)	>1950[b] (13.5)[b]	50 (0.3)	200 (1.4)	100 (0.7)
Impact modifier	9% Novalene EPDM	50 (0.3)	150 (1.0)	>1650[b] (>11.4)[b]	200 (1.4)	200 (1.4)	100 (0.7)
Flame retardant	9% PE-68 4% Antimony oxide	50 (0.3)	50 (0.3)	>1950[b] (>13.5)[b]	50 (0.3)	200 (1.4)	250 (1.7)
Smoke suppressant	13% Firebrake ZB	50 (0.3)	50 (0.3)	>1950[b] (>13.5)[b]	50 (0.3)	200 (1.4)	100 (0.7)
Lubricant	0.1% Calcium stearate 24-26	50 (0.3)	50 (0.3)	>1950[b] (>13.5)[b]	50 (0.3)	200 (1.4)	100 (0.7)
Filler	20% Cimpact 600 Talc	50 (0.3)	50 (0.3)	>1950[b] (>13.5)[b]	100 (0.7)	200 (1.4)	100 (0.7)
Colorant	0.1% Watchung Red RT-428-D	50 (0.3)	50 (0.3)	>1950[b] (>13.5)[b]	50 (0.3)	200 (1.4)	100 (0.7)
Antistatic	0.2% Armostat 475	50 (0.3)	200 (1.4)	>1950[b] (>13.5)[b]	200 (1.4)	200 (1.4)	100 (0.7)

[a] All testing was done according to the block shear method (ASTM D4501).
[b] Due to the severe deformation of the block shear specimens, testing was stopped before the acutal bond strength achieved by the adhesive could be determined (the adhesive bond never failed).

Table 18.30 Shear strength of PP to PP in psi (MPa)

number of reassemblies are possible; repeated removal and reassembly may cause new threads to be cut over the original thread, resulting in a stripped thread. For repeated assembly, the screw should be carefully inserted into the original thread by hand. As shown in Figure 18.74, different self-tapping screws are used. The Hi Lo is a double-lead screw, available as either thread-forming or thread-cutting versions; the Trilobe is a thread-forming screw with triangular threads.

INSERT

A very viable assembly method is using inserts. Inserts are plastics or metals that are inserted into another material (Fig. 18.75). They are commonly used for applications that require frequent disassembly or when a limited engagement length prevents the use of screws. Inserts are used in structural foam and other materials with low shear strengths that cannot withstand fastener loads alone. Threaded inserts are available in internal thread sizes from 4–40 to ¼–20 (in) or M2–M6 (metric). Various

Plastic Material Composition (Occidental Chemical Oxychem 160)		Loctite Adhesive					
		Black Max 380 rubber toughened cyanoacrylate (200 cP)	Prism 401 surface insensitive ethyl cyanoacrylate (100 cP)	Prism 401/ Prism Primer 770 polyolefin primer for cyanoacrylate	Super Bonder 414 general purpose cyanoacrylate (110 cP)	Depend 330 two-part no-mix acrylic	Loctite 3105 light cure acrylic (300 cP)
Unfilled resin	3 rms	>1600[a] (>11.0)[a]	>3650[a] (>25.2)[a]	>2850[a] (>19.7)[a]	>2900[a] (>20.0)[a]	>2650[a] (>18.3)[a]	>2550[a] (>17.6)[a]
Roughened	27 rms	>1600[a] (>11.0)[a]	>1850[a] (>12.8)[a]	>1400[a] (>9.7)[a]	>2900[a] (>20.0)[a]	>1550[a] (>10.7)[a]	>2550 (>17.6)
UV stabilizer	1% UV-531	>1600[a] (>11.0)[a]	>2800[a] (>19.3)[a]	>1400[a] (>9.7)[a]	>2900[a] (>20.0)[a]	>1850[a] (>12.8)[a]	>2550[a] (>17.6)[a]
Impact modifier	7% Paraloid BTA753	>1100[a] (>7.6)[a]	>4300[a] (>29.7)[a]	>3650[a] (>25.2)[a]	>2900[a] (>20.0)[a]	>1050 (7.24)	>3000[a] (>20.7)[a]
Flame retardant	0.3% Antimony Oxide	>1600[a] (>11.0)[a]	>3050[a] (>21.0)[a]	>2850[a] (>19.7)[a]	>2900[a] (>20.0)[a]	>2050[a] (>14.1)[a]	>2550[a] (>17.6)[a]
Smoke suppressant	0.3% Ammonium Octamolybdate	1250 (8.6)	>3650[a] (>25.2)[a]	>2850[a] (>19.7)[a]	>2900[a] (>20.0)[a]	>1800 (>12.4)[a]	>2550[a] (>17.6)[a]
Lubricant	1% Calcium Stearate 24-46	>1600[a] (>11.0)[a]	>3650[a] (>25.2)[a]	>2850[a] (19.7)[a]	>2900[a] (>20.0)[a]	>1900[a] (>13.1)[a]	>2550[a] (>17.6)[a]
Filler	9% OmyaCarb F	>1600[a] (>11.0)[a]	>4250[a] (>29.3)[a]	>1750[a] (>12.1)[a]	>4400[a] (>30.3)[a]	>2650[a] (>18.3)[a]	>3150[a] (>21.7)[a]
Plasticizer	5% Drapex 6.8	>1600[a] (>11.0)[a]	>2250[a] (>15.5)[a]	>1550[a] (>10.7)[a]	>2900[a] (>20.0)[a]	>1500[a] (>10.3)[a]	>2550[a] (>17.6)[a]
Colorant	0.5% FD&C Blue #1	>1600[a] (>11.0)[a]	>3650[a] (>25.2)[a]	>2850[a] (>19.7)[a]	>2900[a] (>20.0)[a]	>1050[a] (>7.24)[a]	>2550[a] (>17.6)[a]
Antistatic	1.5% Markstat AL48	>1600[a] (>11.0)[a]	>3650[a] (>25.2)[a]	>1200[a] (>8.3)[a]	>2900[a] (>20.0)[a]	>900[a] (>6.2)[a]	>2550[a] (>17.6)[a]

[a] The force applied to the test specimens exceeded the strength of the material resulting in substrate failure before the actual bond strength achieved by the adhesive could be determined.
[b] All testing was done according to the block shear method (ASTM D4501).

Table 18.31 Shear strength of PVC to PVC in psi (MPa)

lengths are available in each size for different depths. The performance of an insert is dependent on the shear strength of the material and the knurl pattern on the outer surface of the insert.

Molding inserts directly into the part during a molding process is a convenient and popular method of insert application. Molded-in inserts provide an anchor for machine screws and are commonly made from metals such as steel, brass, and aluminum. Inserts are placed on core pins in the mold, and the plastic part is then molded around them in the mold cavity. Molding in inserts does not require a secondary insertion operation and provides higher torque and pullout properties than other insertion methods; however, high stresses are created in the areas around the inserts.

Molded-in inserts are not recommended for use with some plastics because of the high stresses they produce. Molded-in inserts perform better in higher-creep, crystalline polyolefins than with rigid amorphous resins (chapters 1 and 19). Molded-in stresses result in high failure rates in the form of cracking in low-creep materials.

Plastic Material Composition (Dow Chemical Calibre 300-4)		Loctite Adhesive					
		Black Max 380 rubber toughened cyanoacrylate (200 cP)	Prism 401 surface insensitive ethyl cyanoacrylate (100 cP)	Prism 401/ Prism Primer 770 polyolefin primer for cyanoacrylate	Super Bonder 414 general purpose cyanoacrylate (110 cP)	Depend 330 two-part no-mix acrylic	Loctite 3105 light cure acrylic (300 cP)
Unfilled resin	3 rms	750 (5.2)	3850 (26.6)	2000 (13.8)	1600 (11.0)	1100 (7.6)	3700 (25.5)
Roughened	18 rms	1600 (11.0)	4500 (31.0)	3400 (23.5)	3950 (27.2)	1100 (7.6)	4550 (31.4)
Antioxidant	0.1% Irgafos 168 0.1% Irganox 1076	750 (5.2)	3850 (26.6)	2000 (13.8)	3950 (27.2)	550 (3.8)	3700 (25.5)
UV stabilizer	0.4% Tinuvin 234	750 (5.2)	3850 (28.6)	2000 (13.8)	1600 (11.0)	450 (3.1)	3700 (25.5)
Flame retardant	2% BT-93 1% Antimony oxide	1300 (9.0)	>4100[a] (>28.3)[a]	>3800[a] (>26.2)[a]	>3400[a] (>23.5)[a]	300 (2.1)	3700 (25.5)
Impact modifier	5% Paraloid EXL3607	1000 (6.9)	3850 (26.6)	2000 (13.0)	>4500[a] (>31.0)[a]	500 (3.5)	3700 (25.5)
Lubricant	0.3% Mold Wiz INT-33UDK	1300 (9.0)	3850 (26.6)	2000 (13.8)	3850 (26.6)	1100 (7.6)	3700 (25.5)
Glass filler	23% Type 3090 glass fiber	1150 (7.9)	3850 (26.6)	600 (4.1)	2700 (18.6)	1100 (7.6)	4850 (33.5)
Colorant	4% CPC07327	1650 (11.4)	3850 (26.6)	500 (3.5)	3950 (27.2)	1100 (7.6)	3700 (25.5)

[a] The force applied to the test specimens exceeded the strength of the material resulting in substrate failure before the actual bond strength achieved by the adhesive could be determined.
[b] All testing was done according to the block shear method (ASTM D4501).

Table 18.32 Shear strength of polycarbonate (PC) to PC in psi (MPa)

Inserts are also put in TPs after products are fabricated (Fig. 18.76). They are heated before insertion into the plastic material. The heated insert softens the plastic, which then flows around the outer surface of the insert. Thermal heating can occur by direct contact with a hot body or by preheating in a temperature-controlled chamber. Installation equipment is inexpensive, but insertion is slow.

Press Fit

In press or interference fits, a shaft of one material is joined with the hub of another material by a dimensional interference between the shaft's outside diameter and the hub's inside diameter. Press fitting is an economical procedure that requires only simple tooling, but it produces very high stresses in the plastic parts. It can be used to join parts of the same material as well as dissimilar materials.

Press fits that depend on having a mechanical interface provide a fast and clean assembly. A common usage is to have a plastic hub or boss that accepts either a plastic or metal shaft or pin. The press-fit procedure tends to expand the hub, creating tensile or hoop stress. If the interference is too great, a high strain and stress will develop. The plastic product will do one of several things. It may fail immediately by developing a crack parallel to the axis of the hub to relieve the stress, which

Plastic Material Composition (Dow Chemical Pellethane 2363-55D)		Loctite Adhesive					
		Black Max 380 rubber toughened cyanoacrylate (200 cP)	Prism 401 surface insensitive ethyl cyanoacrylate (100 cP)	Prism 401/ Prism Primer 770 polyolefin primer for cyanoacrylate	Super Bonder 414 general purpose cyanoacrylate (110 cP)	Depend 330 two-part no-mix acrylic	Loctite 3105 light cure acrylic (300 cP)
Unfilled resin (shore D)	14 rms	200 (1.4)	350 (2.4)	1400 (9.7)	300 (2.1)	350 (2.4)	1150 (7.9)
Roughened	167 rms	350 (2.4)	1350 (9.3)	1950 (13.5)	1300 (9.0)	1500 (10.3)	1700 (11.7)
UV stabilizer	1% Tinuvin 328	100 (0.7)	200 (1.4)	950 (6.6)	150 (1.0)	350 (2.4)	750 (5.2)
Flame retardant	15% BT-93 2% Antimony Oxide	200 (1.4)	450 (3.1)	>1850[b] (>12.8)[b]	600 (4.1)	>1400[b] (>9.7)[b]	>1350[b] (>9.3)[b]
Plasticizer	13% TP-95	50 (0.3)	150 (1.0)	>750[b] (>5.2)[b]	150 (1.0)	200 (1.4)	450 (3.1)
Lubricant #1	0.5% Mold Wiz INT-33PA	200 (1.4)	800 (5.5)	>2150[b] (>14.8)[b]	700 (4.8)	900 (6.2)	>1800[b] (>12.4)[b]
Lubricant #2	0.5% FS1235 Silicone	450 (3.1)	>2250[b] (>15.5)[b]	>2900[b] (>20.0)[b]	1250 (8.6)	>2650[b] (>18.3)[b]	>2350[b] (>16.2)[b]
Unfilled resin (shore A)	Estane 58630 B.F. Goodrich	200 (1.4)	>850[b] (>5.9)[b]	>1300[b] (>9.0)[b]	550 (3.8)	450 (3.1)	800 (5.5)

[a] All testing was done according to the block shear method (ASTM D4501).
[b] Due to the severe deformation of the block shear specimens, testing was stopped before the actual bond strength achieved by the adhesive could be determined (the adhesive bond never failed).

Table 18.33 Shear strength of PUR to PUR in psi (MPa)

is a typical hoop-stress failure. It could survive the assembly process, but fail prematurely in use, for a variety of reasons related to its high induced-stress levels. Or it might undergo stress relaxation sufficient to reduce the stress to a lower level that can be maintained

Hoop-stress equations for press-fit situations are used (Fig. 18.77). The allowable design stress or strain will depend on the particular plastic, the temperature, and other environmental considerations. Hoop stress can be obtained by multiplying the appropriate modulus. For high strains, the secant modulus will give the initial stress; the apparent or creep modulus should be used for more long-term stresses. The maximum strain or stress must be below the value that will produce creep rupture in the material. There could be a weld line in the hub that can significantly affect the creep-rupture strength of most plastics.

Complications could develop during processing with press fits in that a round hub or boss may not be the correct shape. Strict processing controls are used to eliminate these types of potential problems. There is a tendency for a round hub to be slightly elliptical in cross-section, increasing the stresses on the part. For critical product performance and in view of what could occur, life-type prototyping testing should be conducted under actual conditions in critical applications.

The consequences of stress will depend on many factors, such as the temperature during and after the assembly of the press fit, the modulus of the mating material, the type of stress, and the environment in which the product is used. The most important factor, though is the type of material

Plastic Material Composition (Allied-Signal Capron 8202 Nylon 6)		Loctite Adhesive					
		Black Max 380 rubber toughened cyanoacrylate (200 cP)	Prism 401 surface insensitive ethyl cyanoacrylate (100 cP)	Prism 401/ Prism Primer 770 polyolefin primer for cyanoacrylate	Super Bonder 414 general purpose cyanoacrylate (110 cP)	Depend 330 two-part no-mix acrylic	Loctite 3105 light cure acrylic (300 cP)
Unfilled resin	11 rms	2450 (16.9)	4500 (31.0)	1600 (11.0)	4100 (28.3)	450 (3.1)	1400 (9.7)
Roughened	15 rms	2450 (16.9)	4500 (31.0)	1600 (11.0)	4100 (28.3)	450 (3.1)	1400 (9.7)
Antioxidant	0.35% Irganox B1171	2450 (16.9)	4500 (31.0)	250 (1.7)	4100 (28.3)	450 (3.1)	1400 (9.7)
UV stabilizer	0.63% Chimasorb 944	2450 (16.9)	4500 (31.0)	1600 (11.0)	4100 (28.3)	450 (3.1)	1400 (9.7)
Impact modifier	5% EXL 3607	>2200[a] (>15.2)[a]	>4500[a] (>31.0)[a]	>1600[a] (>11.0)[a]	>4300[a] (>29.7)[a]	450 (3.1)	1400 (9.7)
Flame retardant	18% PO-64P 4% Antimony Oxide	1700 (11.0)	4500 (31.0)	1600 (11.0)	4100 (28.3)	450 (3.1)	1400 (9.7)
Lubricant #1	0.5% Aluminum Stearate	1450 (10.0)	4500 (31.0)	350 (2.4)	4600 (31.7)	450 (3.1)	1400 (9.7)
Lubricant #2	0.5% Moldwiz INT-33PA	2450 (16.9)	>4500[a] (>31.0)[a]	550 (3.8)	>3750[a] (>25.9)[a]	450 (3.1)	1050 (7.2)
Glass filler	30% Type 3450 Glass Fiber	2450 (16.9)	>4700[a] (>32.4)[a]	150 (1.0)	>4450[a] (>30.7)[a]	450 (3.1)	1400 (9.7)
Talc filler	30% Mistron CB Talc	2450 (16.9)	2200 (15.2)	2100 (14.5)	2750 (19.0)	450 (3.1)	1400 (9.7)
Plasticizer	4% Ketjen-Flex 8450	3300 (22.8)	>4550[a] (>31.4)[a]	650 (4.5)	>4450[a] (>30.7)[a]	450 (3.1)	1400 (9.7)
Antistatic	5% Larostat HTS 906	2450 (16.9)	>3100[a] (>21.4)[a]	350 (2.4)	>4100[a] (>28.3)[a]	450 (3.1)	1400 (9.7)

[a] The force applied to the test specimens exceeded the strength of the material resulting in substrate failure before the actual bond strength achieved by the adhesive could be determined.
[b] All testing was done according to the block shear method (ASTM D4501).

Table 18.34 Shear strength of PA to PA in psi (MPa)

being used. Some substances will creep, or stress may relax, but other substances will fracture or craze if the strain is too high. Except for light press fits, this type of assembly design can be risky for the novice because a weld line might already weaken the boss.

Molded-in inserts are usually used to develop good holding power between the insert and the molded plastic (the previous section reviewed this subject). A guide for the wall thickness around the insert is given in Table 4.17 in volume 1. See also Table 18.42.

SNAP-FIT

In snap-fit fastening, two parts are mechanically joined through an interlocking configuration that is molded into the parts. Many different configurations are possible to accommodate different part

Plastic Material Composition (DuPont Polymers Vespel and Kapton)		Loctite Adhesive					
		Black Max 380 rubber toughened cyanoacrylate (200 cP)	Prism 401 surface insensitive ethyl cyanoacrylate (100 cP)	Prism 401/ Prism Primer 770 polyolefin primer for cyanoacrylate	Super Bonder 414 general purpose cyanoacrylate (110 cP)	Depend 330 two-part no-mix acrylic	Loctite 3105 light cure acrylic (300 cP)
Vespel SP-1	unfilled	1550 (10.7)	2200 (15.2)	350 (2.4)	1650 (11.4)	1150 (7.9)	800 (5.5)
Vespel SP-21	15% graphite	1400 (9.7)	2250 (15.5)	850 (5.9)	2350 (16.2)	550 (3.8)	1000 (6.9)
Vespel SP-22	40% graphite 10% PTFE	550 (3.8)	850 (5.9)	400 (2.8)	1000 (6.9)	500 (3.5)	250 (1.7)
Vespel SP-211	15% graphite 10% PTFE	400 (2.8)	550 (3.8)	600 (4.1)	700 (4.8)	200 (1.4)	200 (1.4)
Kapton HN	5 mil thick 500 gauge film	>800[a,b,c] (>5.5)[a,b,c]	>800[a,c] (>5.5)[a,c]	650[c] (4.5)[c]	>800[a,c] (>5.5)[a,c]	>800[a,c] (>5.5)[a,c]	>800[a,c] (>5.5)[a,c]
Kapton HPP-ST	5 mil thick 500 gauge film	>800[b,c] (>5.5)[b,c]	>800[a,c] (>5.5)[a,c]	600[c] (4.1)[c]	>800[a,c] (>5.5)[a,c]	>800[a,c] (>5.5)[a,c]	>800[a,c] (>5.5)[a,c]
Kapton HPP-FST	5 mil thick 500 gauge film	>800[a,c] (>5.5)[a,c]	>800[a,c] (>5.5)[a,c]	450[c] (3.1)[c]	>800[a,c] (>5.5)[a,c]	>800[a,c] (>5.5)[a,c]	>800[a,c] (>5.5)[a,c]

[a] The force applied to the test specimens exceeded the strength of the material resulting in substrate failure before the actual bond strength achieved by the adhesive could be determined.
[b] TAK PAK 7452 Accelerator was used in conjunction with Black Max 380.
[c] The Kapton films were bonded to aluminum lap shears prior to evaluation.
[d] All testing was done according to the block shear method (ASTM D4501).

Table 18.35 Shear strength of polyimide to polyimide in psi (MPa)

designs. In snap-fits, a protrusion on one part (a hook, a stud, or a bead, for example) is briefly deflected during joining to catch in a depression or undercut molded into the other part. The force required for joining depends on the snap-fit design. After the brief joining stress, the joint is vibration resistant and usually stress-free.

Snap-fits are used in a variety of industries to assemble power tools, computer cases, electronic components, toys, automobile parts, medical devices, washing machines, pens, bottles, and packaging boxes. Snap-fits can be used as a temporary holder for other assembly methods, such as adhesive bonding or welding.

Snap-fitting is an economical, rapid, and a very popular assembly method. It is used to join two dissimilar plastics or plastics to metals, and can snap-fits can be designed for permanent fastening or for repeated disassembly. Hermetic or moisture-resistant seals are possible in some designs. Snap-fits require more attention to engineering design than other mechanical fastening methods and can fail before or during assembly or during use if not designed properly (chapter 19). Stress analysis of some snap-fit designs can be performed using hand calculations; designs with more complicated geometries may require finite element analysis for accurate results.

The most common type of snap-fit is the cantilever beam. A cantilever beam snap-fit is a hook and groove joint in which a protrusion from one part interlocks with a groove on the other part (Fig. 18.78). Cantilever beam snap-fits can be straight or may have a bend in the beam (curved

Plastic Material Composition (Du Pont Delrin 100)		Loctite Adhesive					
		Black Max 380 rubber toughened cyanoacrylate (200 cP)	Prism 401 surface insensitive ethyl cyanoacrylate (100 cP)	Prism 401/ Prism Primer 770 polyolefin primer for cyanoacrylate	Super Bonder 414 general purpose cyanoacrylate (110 cP)	Depend 330 two-part no-mix acrylic	Loctite 3105 light cure acrylic (300 cP)
Unfilled Resin	30 rms	100 (0.7)	200 (1.4)	1700 (11.7)	500 (3.5)	50 (0.3)	250 (1.7)
Roughened	47 rms	150 (1.0)	600 (4.1)	1700 (11.7)	500 (3.5)	100 (0.7)	250 (1.7)
Antioxidant	0.2% Irganox 1010	100 (0.7)	400 (2.8)	1700 (11.7)	500 (3.5)	50 (0.3)	250 (1.7)
UV stabilizer	0.2% Tinuvin 328 0.4% Tinuvin 770	100 (0.7)	900 (6.2)	1700 (11.7)	500 (3.5)	50 (0.3)	300 (2.1)
Impact Modifier	30% Estane 5708F1	100 (0.7)	350 (2.4)	1700 (11.7)	500 (3.5)	50 (0.3)	350 (2.4)
Lubricant	0.88% N,N'-Ethylene bisstearamide wax	100 (0.7)	350 (2.4)	1700 (11.7)	900 (6.2)	50 (0.3)	450 (3.1)
Glass filler	20% type 3090 glass fiber	100 (0.7)	1100 (7.6)	2800 (11.7)	1100 (7.6)	50 (0.3)	300 (2.1)
PTFE filler	15% PTFE MP1300	100 (0.7)	200 (1.4)	1700 (11.7)	100 (0.7)	50 (0.3)	250 (1.7)
Colorant	4% 3972 colorant	100 (0.7)	200 (1.4)	1700 (11.7)	500 (3.5)	50 (0.3)	250 (1.7)
Antistatic	1.5% Markstat AL12	150 (1.0)	1750 (12.1)	1700 (11.7)	1100 (7.6)	50 (0.3)	250 (1.7)
Acetal copolymer	Celcon courtesy of Hoechst Celanese	50 (0.3)	100 (0.7)	300 (2.1)	100 (0.7)	200 (1.4)	200 (1.4)

a All testing was done according to the block shear method (ASTM D4501).

Table 18.36 Shear strength of acetal to acetal in psi (MPa)

beam). Rectangular cross-sections are common; beam cross-sections may also be square, round (hollow or filled), trapezoidal, triangular, convex, or concave (chapter 20). The beam can be of constant width and height or can be tapered to avoid stress concentration near the point of attachment with the part wall.

They undergo flexural stress during assembly and are modeled as cantilever beams in design calculations. After assembly, joints are usually stress-free; however, joints can be designed to be partially loaded after assembly for an extra-tight fit. Loaded snap-fits may be subject to creep or stress relaxation. Other types of snap-fits include annular, used to join spherical parts; torsional, in which a latch is attached to a torsion bar or shaft; ball-and-socket snap-fits, used to transmit motion; and U-shaped snap-fits, commonly used for lid fasteners. Combinations of different types are also possible in one design.

Some design considerations in snap-fits include the forces required for assembly (and disassembly, if required), ease of molding and assembly, the material strain produced during assembly, and other requirements of the application. Stress analysis, based on a geometric model for the particular type of snap-fit, is performed to determine assembly forces, deflections, and stresses produced during assembly.

Plastic Material Composition (ICI Acrylics Inc. Perspex CP80)		Loctite Adhesive					
		Black Max 380 rubber toughened cyanoacrylate (200 cP)	Prism 401 surface insensitive ethyl cyanoacrylate (100 cP)	Prism 401/ Prism Primer 770 polyolefin primer for cyanoacrylate	Super Bonder 414 general purpose cyanoacrylate (110 cP)	Depend 330 two-part no-mix acrylic	Loctite 3105 light cure acrylic (300 cP)
Unfilled resin	3 rms	600 (4.1)	>3950[a] (>27.2)[a]	250 (1.7)	>2900[a] (>20.0)[a]	1150 (7.9)	1750 (12.1)
Roughened	34 rms	1500 (10.3)	2150 (14.8)	400 (2.8)	>2900[a] (>20.0)[a]	1150 (7.9)	1750 (12.1)
Antioxidant	0.1% Irganox 245	1400 (10.0)	>3950[a] (>27.2)[a]	350 (2.4)	>2900[a] (>20.0)[a]	1150 (7.9)	1750 (12.1)
UV stabilizer	0.6% Uvinul 3039	1450 (10.0)	>3950[a] (>27.2)[a]	250 (1.7)	>2900[a] (>20.0)[a]	1150 (7.9)	1750 (12.1)
Flame retardant	17% Phoschek P-30	1050 (7.2)	>5050[a] (>34.8)[a]	>5250[a] (>36.2)[a]	>2900[a] (>20.0)[a]	1150 (7.9)	1750 (12.1)
Lubricant	5% Witconol NP-330	>3050[a] (>21.0)[a]	>3950[a] (>27.2)[a]	350 (2.4)	>4550[a] (>31.4)[a]	1150 (7.9)	1250 (8.6)
Impact modifier	29% Paraloid EXL 3330	1250 (8.6)	>3950[a] (>27.2)[a]	1250 (8.6)	2900 (20.0)	650 (4.5)	1750 (12.1)
Plasticizer	9% Benoflex 50	600 (4.1)	>3000[a] (>20.7)[a]	250 (1.7)	>2900[a] (>20.0)[a]	1150 (7.9)	1750 (12.1)
Colorant A	1% OmniColor Pacific Blue	1550 (10.7)	>3350[a] (>23.1)[a]	250 (1.7)	>2900[a] >20.0)[a]	1150 (7.9)	1350 (9.3)
Colorant B	0.5% 99-41-042	600 (4.1)	>2350[a] (>16.2)[a]	250 (1.7)	>2900[a] (>20.0)[a]	450 (3.1)	1750 (12.1)
Antistatic	1.5% Markstat AL-48	>2150[a] (>14.8 a)	>3950[a] (>27.2)[a]	250 (1.7)	>2900[a] (>20.0)[a]	1150 (7.9)	1750 (12.1)

[a] The force applied to the test specimens exceeded the strength of the material resulting in substrate failure before the actual bond strength achieved by the adhesive could be determined.
[b] All testing was done according to the block shear method (ASTM D4501).

Table 18.37 Shear strength of polymethyl methacrylate (PMMA) to PMMA in psi (MPa)

STAKING

In staking, a head is formed on a TP stud by cold flow or melting of the plastic. The stud protrudes through a hole in the parts being joined, and staking the stud mechanically locks the two parts together (Figs. 18.79 and 18.80).

Staking can be performed by four different methods: ultrasonic staking, cold staking, heat staking, and thermostaking. Cold staking, or heading, uses high pressures to induce cold flow of the plastic material; pressures of at least 6000 psi (41 MPa) are generally required. Stud lengths are approximately 1.5 times the stud diameter; stud length includes part thickness. Because cold heading creates high stresses on the stud, only more malleable TPs are suitable for this process. Soft, brittle, or fragile materials are not usually assembled by cold staking.

In heat staking, heated probes and low to moderate pressure are used to compress and reform the stud. Because stresses are lower than in cold staking, heat staking can be used to join a variety of plastic materials. Heat staking is used to join two parts of the same plastic or dissimilar plastics. It is an economical process that produces consistent results.

Plastic Material Composition (Hoechst Celanese Resin T80)		Loctite Adhesive					
		Black Max 380 rubber toughened cyanoacrylate (200 cP)	Prism 401 surface insensitive ethyl cyanoacrylate (100 cP)	Prism 401/ Prism Primer 770 polyolefin primer for cyanoacrylate	Super Bonder 414 general purpose cyanoacrylate (110 cP)	Depend 330 two-part no-mix acrylic	Loctite 3105 light cure acrylic (300 cP)
Unfilled resin	7 rms	450 (3.1)	>3200[a] (>22.1)[a]	>1800[a] (>12.4)[a]	>2200[a] (>15.2)[a]	500 (3.5)	1150 (7.9)
Roughened	31 rms	200 (1.4)	900 (6.2)	700 (4.8)	950 (6.6)	500 (3.5)	1150 (7.9)
Impact modifier	17% Novalene 7300P	>250[a] (>1.7)[a]	>350[a] (>2.4)[a]	<250[a] (>1.7)[a]	>400[a] (>2.8)[a]	>150[a] (>1.0)[a]	>300[a] (>2.1)[a]
Flame retardant	15% PO-64P 4% Antimony Oxide	1550 (10.7)	>2150[a] (>14.8)[a]	600 (4.14)	>2200[a] (>15.2)[a]	850 (5.9)	1150 (7.9)
Lubricant	0.2% Zinc Stearate	750 (5.2)	>1800[a] (>12.4)[a]	>1800[a] (>12.4)[a]	>2200[a] (>15.2)[a]	500 (3.5)	1150 (7.9)
Internal mold release	0.5% Mold Wiz 33PA	800 (5.5)	>3200[a] (>22.1)[a]	>1800[a] (>12.4)[a]	>2200[a] (>15.2)[a]	500 (3.5)	1700 (11.7)
Filler	17% 3540 Fiberglass	800 (5.5)	2900 (20.0)	>3350[a] (>23.1)[a]	2200 (15.2)	800 (5.5)	1700 (11.7)
Colorant	0.5% Green 99-41042	1000 (6.9)	>2200[a] (>15.2)[a]	>1800[a] (>12.4)[a]	>2200[a] (>15.2)[a]	500 (3.5)	1150 (7.9)
Antistatic	1% Dehydat 8312	>1350[a] (>9.3)[a]	>1900[a] (>13.1)[a]	>1800[a] (>12.4)[a]	>1450[a] (>10.0)[a]	500 (3.5)	1150 (7.9)

[a] The force applied to the test specimens exceeded the strength of the material resulting in substrate failure before the actual bond strength achieved by the adhesive could be determined.
[b] All testing was done according to the block shear method (ASTM D4501).

Table 18.38 Shear strength of polyethylene terephthalate (PET) to PET in psi (MPa)

Plastic Material Composition		Loctite Adhesive					
		Black Max 380 rubber toughened cyanoacrylate (200 cP)	Prism 401 surface insensitive ethyl cyanoacrylate (100 cP)	Prism 401/ Prism Primer 770 polyolefin primer for cyanoacrylate	Super Bonder 414 general purpose cyanoacrylate (110 cP)	Depend 330 two-part no-mix acrylic	Loctite 3105 light cure acrylic (300 cP)
Victrex 450G control	unfilled resin courtesy of Victrex, USA 4 rms	150 (1.0)	250 (1.7)	250 (1.7)	200 (1.4)	350 (2.4)	1100 (7.6)
450G roughened	22 rms	700 (4.8)	350 (2.4)	350 (2.4)	300 (2.1)	350 (2.4)	1100 (7.6)
PEEK 450 CA30	30% carbon fiber courtesy of Victrex, USA	150 (1.0)	200 (1.4)	450 (3.1)	250 (1.7)	450 (3.1)	950 (6.6)
Thermocomp LF-1006	30% glass fiber courtesy of LNP Engineering Plastics	100 (0.7)	250 (1.7)	550 (3.8)	400 (2.8)	500 (3.5)	1200 (8.3)
Lubricomp LCL-4033 EM	15% carbon fiber, 15% PTFE courtesy of LNP Engineering Plastics	100 (0.7)	400 (2.8)	300 (2.1)	250 (1.7)	500 (3.5)	900 (6.2)

[a] All testing was done according to the block shear method (ASTM D4501).

Table 18.39 Shear strength of polyetheretherketone (PEEK) to PEEK in psi (MPa)

Plastic Material Composition (Amoco Performance Products Xydar)		Loctite Adhesive					
		Black Max 380 rubber toughened cyanoacrylate (200 cP)	Prism 401 surface insensitive ethyl cyanoacrylate (100 cP)	Prism 401/ Prism Primer 770 polyolefin primer for cyanoacrylate	Super Bonder 414 general purpose cyanoacrylate (110 cP)	Depend 330 two-part no-mix acrylic	Loctite 3105 light cure acrylic (300 cP)
G-540	40% glass reinforced 63 rms	500 (3.5)	300 (2.1)	400 (2.8)	350 (2.4)	450 (3.1)	650 (4.5)
G-540 roughened	58 rms G-930	1050 (7.2)	1100 (7.6)	1050 (7.2)	1100 (7.6)	1150 (7.9)	650 (4.5)
G-930	30% glass reinforced 106 rms	350 (2.4)	300 (2.1)	500 (3.5)	350 (2.4)	500 (3.5)	500 (3.5)
G-930 roughened	G-930 roughened 113 rms	1200 (8.3)	1450 (10.0)	1550 (10.7)	1250 (8.6)	900 (6.2)	500 (3.5)

a All testing was done according to the block shear method (ASTM D4501).

Table 18.40 Shear strength of liquid crystal polymer (LCP) to LCP in psi (MPa)

Plastic Material Composition (DuPont Polymers Teflon)		Loctite Adhesive					
		Black Max 380 rubber toughened cyanoacrylate (200 cP)	Prism 401 surface insensitive ethyl cyanoacrylate (100 cP)	Prism 401/ Prism Primer 770 polyolefin primer for cyanoacrylate	Super Bonder 414 general purpose cyanoacrylate (110 cP)	Depend 330 two-part no-mix acrylic	Loctite 3105 light cure acrylic (300 cP)
Unfilled resin	88 rms	200 (1.4)	350 (2.4)	1050 (7.2)	300 (2.1)	100 (0.7)	150 (1.0)
Roughened	349 rms	200 (1.4)	350 (2.4)	800 (5.5)	700 (4.8)	250 (1.7)	300 (2.1)
Teflon treated with Acton Fluoro Etch		950 (6.6)	1800 (12.4)	1550 (10.7)	1750 (12.1)	450 (3.1)	750 (5.2)
Teflon treated with Gore Tetra Etch		1350 (9.3)	1900 (13.1)	1200 (8.3)	1800 (12.4)	350 (2.4)	700 (4.8)
Ethylene tetrafluoroethylene copolymer (ETFE)		50 (0.3)	100 (0.7)	>1650 (>11.4)	100 (0.7)	50 (0.3)	100 (0.7)
Fluorinated ethylene-propylene (FEP)		<50 (<0.3)	<50 (<0.3)	<50 (<0.3)	<50 (<0.3)	<50 (<0.3)	<50 (<0.3)
Polyperfluoroalkoxyethylene (PFA)		<50 (<0.3)	100 (0.7)	400 (2.8)	50 (0.3)	<50 (<0.3)	50 (0.3)

a All testing was done according to the block shear method (ASTM D4501).

Table 18.41 Shear strength of fluoroplastic to fluoroplastic in psi (MPa)

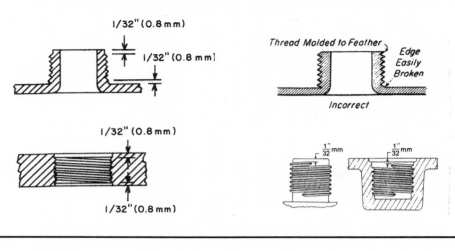
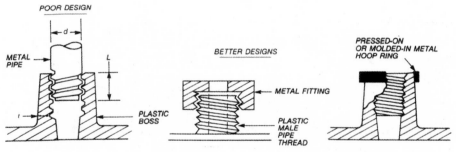
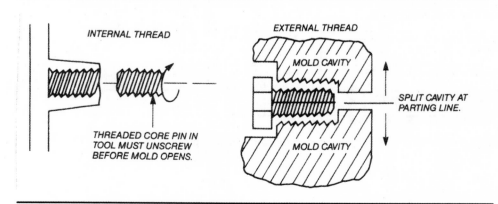

Figure 18.72 Guide for molding threads.

Figure 18.73 Examples of assembling all plastic and plastic to different materials where thermal stresses can become a problem when proper design is not used (chapter 19).

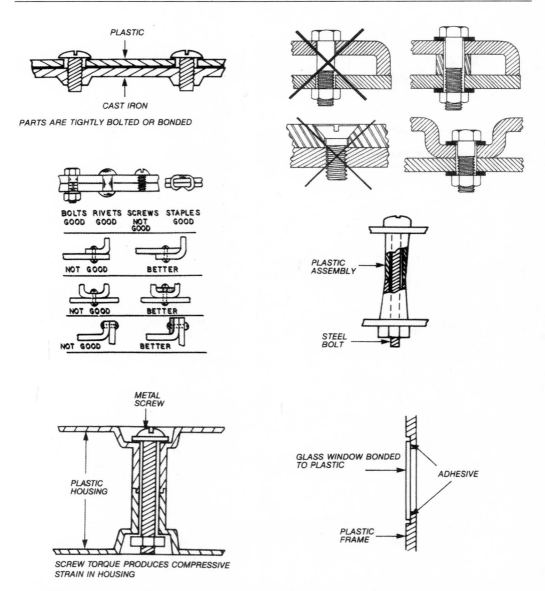

Figure 18.73 Examples of assembling all plastic and plastic to different materials where thermal stresses can become a problem when proper design is not used (chapter 19) *(continued)*.

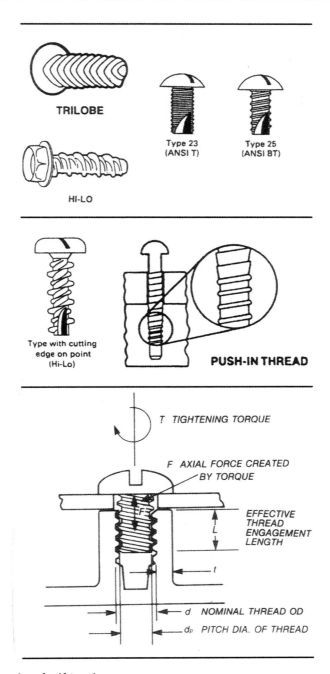

Figure 18.74 Examples of self-tapping screws.

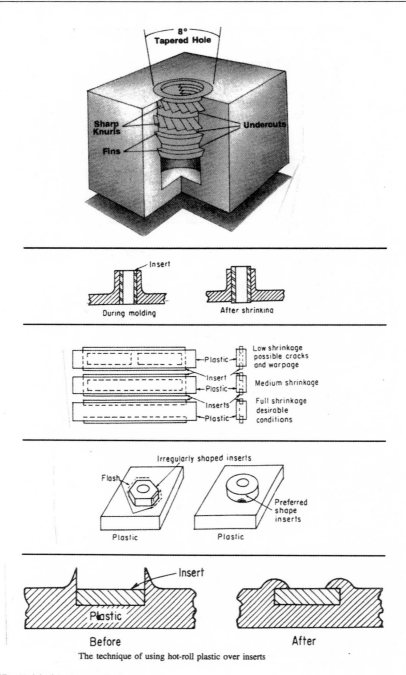

Figure 18.75 Molded-in insert designs.

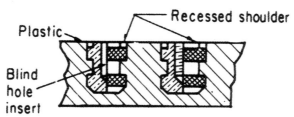

During molding, any loose steel inserts falling into the molds will damage their cavities, but brass or soft metals like aluminum will usually crush, with minimal or no damage to the mold. The flow of plastic into the interior of an insert is impeded by using a blind-hole insert and a shoulder (about $\frac{1}{32} \times \frac{1}{32}$ in.) around the insert opening.

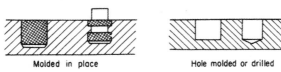

Some of the methods of assembling inserts include mechanical pushing and sonic pushing, or allowing hot plastic to shrink around the insert.

Plastic generally shrinks away from metal when it is molded inside a metal insert. Both the insert's structure and the type of plastic used will determine the amount of shrinkage.

Metal stamping and inserts of various shapes are usable in many ways. To prevent cracking and crazing during aging under use surround all inserts with reasonably thick plastic walls. Thin walls can crack and too-thin walls can also show sink marks.

Figure 18.75 Molded-in insert designs *(continued)*.

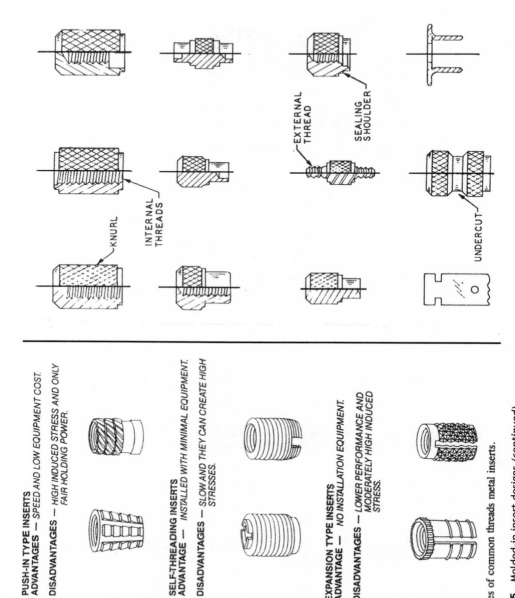

Figure 18.75 Molded-in insert designs (continued).

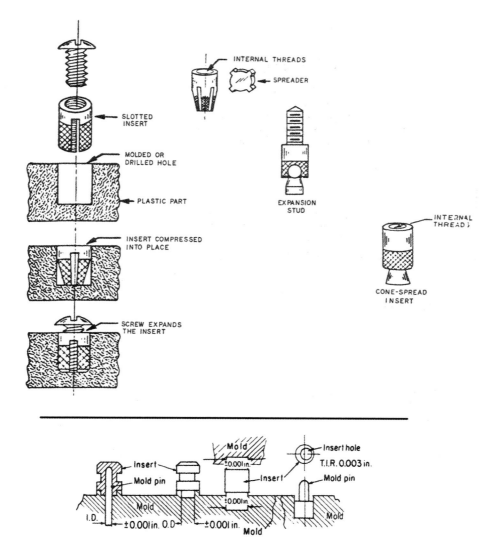

Figure 18.76 Examples of metal-expansion types of slotted and nonslotted inserts.

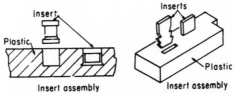

Assembling inserts after molding prevents problems such as potential melt flow over metal surfaces during molding and the scratching of plated inserts during any required deflashing. Pressed-in inserts require holes sized for proper fits.

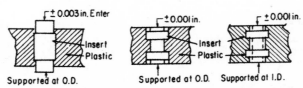

Generally, through inserts must be molded ±0.001 in. of length to ensure their making contact with both mold surfaces.

Knurls and grooves can serve to anchor inserts in molded products. They should be flush with the top surface or in contact with an assembling member to prevent their jacking out.

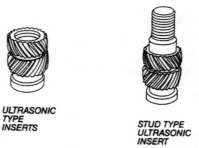

These examples of ultrasonic inserts, designed for excellent performance, result in fast action with very little induced stress.

Figure 18.76 Examples of metal-expansion types of slotted and nonslotted inserts *(continued)*.

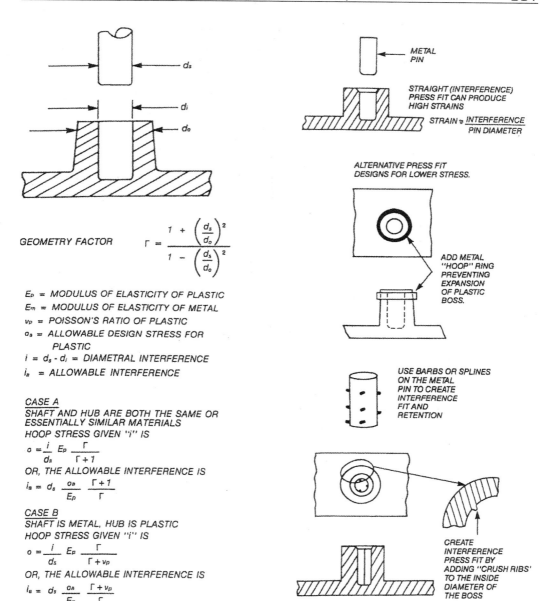

Figure 18.77 Examples of press-fit-stress analyses (courtesy of Bayer).

	Diameter of Inserts, in.					
Plastic Material	.125 (3.17)	.250 (6.35)	.375 (9.52)	.500 (12.7)	.750 (19.0)	1.00 (25.4)
ABS	.125 (3.17)	.250 (6.35)	.375 (9.52)	.500 (12.7)	.750 (19.0)	1.00 (25.4)
Acetal	.062 (1.57)	.125 (3.17)	.187 (4.75)	.250 (6.35)	.375 (9.52)	.500 (12.7)
Acrylics	.093 (2.36)	.125 (3.17)	.187 (4.75)	.250 (6.35)	.375 (9.52)	.500 (12.7)
Cellulosics	.125 (3.17)	.250 (6.35)	.375 (9.52)	.500 (12.7)	.750 (19.0)	1.00 (25.4)
Ethylene vinyl acetate	.040 (1.02)	.085 (2.16)	N.R.	N.R.	N.R.	N.R.
F.E.P. (fluorocarbon)	.025 (0.64)	.060 (1.52)	N.R.	N.R.	N.R.	N.R.
Nylon	.125 (3.17)	.250 (6.35)	.375 (9.52)	.500 (12.7)	.750 (19.0)	1.00 (25.4)
Polycarbonate	.062 (1.57)	.125 (3.17)	.187 (4.75)	.250 (6.35)	.375 (9.52)	.500 (12.7)
Polyethylene (H.D.)	.125 (3.17)	.250 (6.35)	.375 (9.52)	.500 (12.7)	.750 (19.0)	1.00 (25.4)
Polypropylene	.125 (3.17)	.250 (6.35)	.375 (9.52)	.500 (12.7)	.750 (19.0)	1.00 (25.4)
Phenolic G.P.	.093 (2.36)	.156 (3.96)	.187 (4.75)	.218 (5.53)	.312 (7.92)	.343 (8.71)
Phenolic (medium impact)	.078 (1.98)	.140 (3.56)	.156 (3.96)	.203 (5.16)	.281 (7.14)	.312 (7.92)
Phenolic (high impact)	.062 (1.57)	.125 (3.17)	.140 (3.56)	.187 (4.75)	.250 (6.35)	.281 (7.13)
Urea	.093 (2.36)	.156 (3.96)	.187 (4.75)	.218 (5.53)	.312 (7.92)	.343 (8.71)
Melamine	.125 (3.17)	.187 (4.75)	.218 (5.53)	.312 (7.92)	.343 (8.71)	.375 (9.52)
Epoxy	.020 (0.51)	.030 (0.76)	.040 (1.02)	.050 (1.27)	.060 (1.52)	.070 (1.78)
Alkyd	.125 (3.17)	.187 (4.75)	.187 (4.75)	.312 (7.92)	.343 (8.71)	.375 (9.52)
Diallyl phthalate	.125 (3.17)	.187 (4.75)	.250 (6.35)	.312 (7.92)	.343 (8.71)	.375 (9.52)
Polyester (premix)	.093 (2.36)	.125 (3.17)	.140 (3.56)	.187 (4.75)	.250 (6.35)	.281 (7.14)
Polyester T.P.	.062 (1.57)	.125 (3.17)	.187 (4.75)	.250 (6.35)	.375 (9.52)	.375 (9.52)

Table 18.42 Guide relating molded wall thicknesses to insert diameters (in [mm])

Thermo-pneumatic-staking, or thermostaking, uses a heated, hollow tool to deliver a low volume of superheated air to the TP stud. The tool is lowered over the stud, and the hot air rapidly softens the plastic. The hot airflow is then shut off, and a cold stake probe located at the top of the tool descends onto the stud. A stud head is formed, and, after the plastic solidifies and cools, the cold staking probe is retracted.

In ultrasonic staking, a TP stud is melted and reformed to mechanically lock another, usually dissimilar material in place. A TP stud protrudes through a hole in the dissimilar material, usually metal; ultrasonic energy melts the stud, which compresses under pressure from the equipment horn and takes the shape of the horn cavity. After vibrations stop, the horn remains in contact with the stud until it solidifies. Ultrasonic staking works well with soft or amorphous plastics having low melt flows, allowing the head of the rivet to form by both mechanical and thermal mechanisms (chapter 1). With high-melt materials or materials that require high vibrational amplitudes, melt can flow so rapidly that it is ejected out of the horn contour, resulting in an incomplete rivet head.

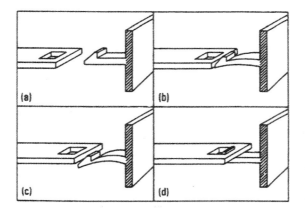

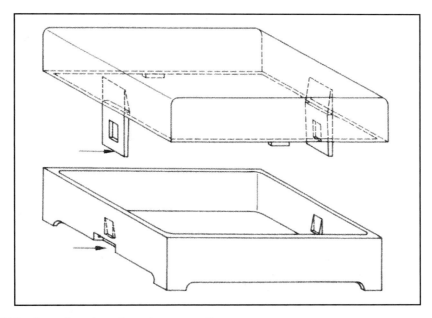

Figure 18.78 Examples of cantilever beam snap-fits.

WELDING ASSEMBLY

Different heat softening methods are used to weld a TP to a TP. An introduction to welding is provided in Tables 18.43 to 18.48. The different processes are used to make permanent bonds between materials that can meet various requirements, such as particular shapes, thickness levels, appearances, different bond strengths, ability to bond different materials, hermetic seals, or effects of

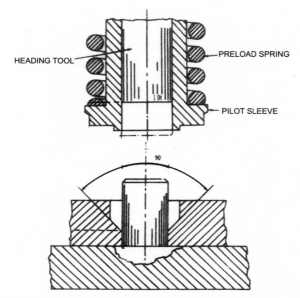

Figure 18.79 Example of cold staking of plastic.

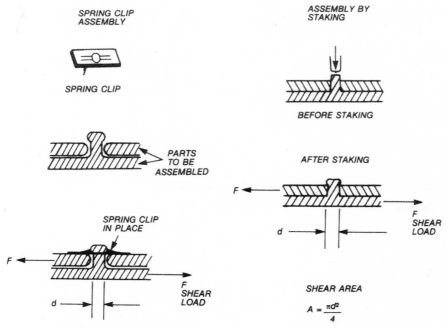

Figure 18.80 Example of hot staking of plastic.

Plastics	Original tensile strength, psi	Hot-air welding, %	Friction welding, %	Hot-plate welding, %	Dielectric welding, %	Polymerization welding, %
Thermoplastics						
Acrylonitrile butadiene styrene	2,400–9,000	50–70	50–70	50–70	50–80	—
Acetal	8,000–10,000	20–30	50–70	20–30	—	—
Cellulose acetate	2,400–8,500	60–75	65–80	65–80	—	—
Cellulose acetate butyrate	3,000–7,000	60–75	65–80	65–80	—	—
Ethyl cellulose	2,000–8,000	50–70	50–70	50–70	—	—
Methyl methacrylate	8,000–11,000	30–70	30–50	20–50	—	60–90
Nylon	7,000–12,000	50–70	50–70	50–70	—	—
Polycarbonate	8,000–9,500	35–50	40–50	40–50	—	—
Polyethylene	800–6,000	60–80	70–90	60–80	—	—
Polypropylene	3,000–6,000	60–80	70–90	60–80	—	—
Polystyrene	3,500–8,000	20–50	30–60	20–50	—	—
Polystyrene acrylonitrile	8,000–11,000	20–60	20–50	20–50	30–50	—
Polyvinyl chloride	5,000–9,000	60–70	50–70	60–70	60–70	—
Saran	3,000–5,000	60–70	50–70	60–70	60–70	—

Table 18.43   Examples of welding methods versus tensile-strength retention

Thermoplastics	Mechanical fasteners	Adhesives	Spin and vibration welding	Thermal welding	Ultrasonic welding	Induction welding	Remarks
ABS	G	G	G	G	G	G	Body type adhesive recommended
Acetal	E	P	G	G	G	G	Surface treatment for adhesives
Acrylic	G	G	F–G	G	G	G	Body type adhesive recommended
Nylon	G	P	G	G	G	G	
Polycarbonate	G	G	G	G	G	G	
Polyester TP	G	F	G	G	G	G	
Polyethylene	P	NR	G	G	G–P	G	Surface treatment for adhesives
Polypropylene	P	P	E	G	G–P	G	Surface treatment for adhesives
Polystyrene	F	G	E	G	E–P	G	Impact grades difficult to bond
Polysulfone	G	G	G	E	E	G	
Polyurethane TP	NR	G	NR	NR	NR	G	
PPO modified	G	G	E	G	G	G	
PVC rigid	F	G	F	G	F	G	

E = Excellent, G = Good, F = Fair, P = Poor, NR = Not recommended.

Table 18.44   Examples of welding characteristics

Material	Percent of Weld Strength[a]	Spot Weld	Staking and Inserting	Swaging	Welding Near-Field[b]	Welding Far-Field[b]
ABS	95–100	E	E	G	E	G
ABS/polycarbonate	95–100	E	E	G	E	G
ABS/PVC	95–100	E	E	G	G	F
Acetal	65–70[e]	G	E	P	G	G
Acrylics	95–100	G	E	P	E	G
Butyrates	90–100	G	G–F	G	P	P
Cellulosics	90–100	G	G–F	G	P	P
Polyethylene	90–100	E	E	G	G–P	F–P
Polypropylene	90–100	E	E	G	G–P	F–P
Acrylic/PVC	95–100	E	E	G	G	F
ASA	95–100	E	E	G	E	G
Methylpentene	90–100	E	E	G	G	F
Modified phenylene oxide	95–100	E	E	F–P	G	E–G
Nylon	90–100	E	E	F–P	G	F
Polyesters (thermoplastic)	90–100	G	G	F	G	F
Phenoxy	90–100	G	E	G	G	G–F
Polyarylsulfone	95–100	G	E	G	E	G
Polycarbonate	95–100	E	E	G–F	E	E
Polyimide	80–90	F	G	P	G	F
Polyphenylene oxide	95–100	E	G	F–P	G	G–F
Polysulfone	95–100	E	E	F	G	G–F
Vinyls	40–100	G	G–F	G	F–P	F–P

E = excellent; G = good; F = fair; P = poor.

Table 18.45 Examples of ultrasonic welding applications

Material	Ultrasonic welding	Linear vibration	Orbital vibration	Hot plate welding	Electromagnetic bonding
Amorphous thermoplastics	1	1	1	1	1
Semicrystalline thermoplastics	2	1	1	1	1
Olefins	2	1	1	1	1
TPRs	1	2	2	2	2
Composites	2	2	2	2	1
Part					
Thin walls	1	3	2	1	2
Complex geometry	2	1	1	1	1
Large parts	2	1	2	1	1
Small parts	1	1	1	1	1
Internal welds	1	2	2	1	1
Long, unsupported walls	1	3	2	1	1

1 = recommended; 2 = limited; 3 = not recommended.

Table 18.46 Comparison of a few welding methods

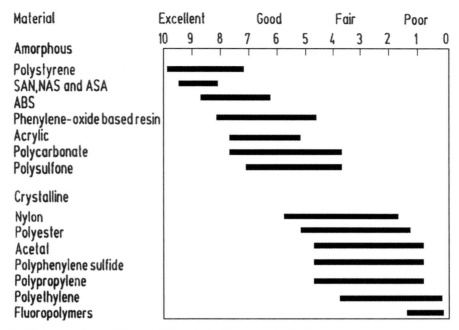

Table 18.47 Comparing welding of different plastics, each to itself

additives or fillers used in the plastics. Once a process is being used, if the compound additives or fillers are changed or added, bond performance can change or even fail. For example, with a certain amount of glass-fiber filler that does not melt, welding action can disappear (7, 249, 579).

Process-related welding equipment has become very important to processors, who want their assembly processes to be documented and in control. Thus there is more demand for higher-level process-control equipment. In the past, device manufacturers could only control one welding variable at a time. Now they also want to control variables such as the energy applied to the part, the distance the part travels, peak power, line pressure, and amplitude. To control each of these variables, manufacturers establish upper and lower limits that allow them to identify and remove parts that do not meet their requirements.

HOT-PLATE WELDING

Hot-plate (tool) welding is a widespread and reliable technique for welding TPs. It is used to join components of variable complexity and size produced by extrusion, injection molding, or others fabricating methods. In cases where the surfaces to be joined are flat, the hot tool is a plate with the temperature controlled on both sides; the process is called flat hot-plate welding (Fig. 18.81 and Tables 18.49 to 18.52).

Process	Equipment cost	Tooling cost	Typical output rates	Normal economic production quantities	Remarks
Ultrasonic welding	Moderately low to high	Moderate to high	1000 pieces per hour, manually loaded	High	Automatic operation possible
Vibration welding	Moderate	Moderate	240 pieces per hour from single cavity, manually loaded	Medium and high	Setup time 10 min; multiple cavities and mechanized loading possible
Spin welding	Moderate	Moderate	640 pieces per hour, manually loaded	High	Setup time $\frac{1}{2}$ h; mechanization possible
Hot-plate welding	Moderately low to high	Moderate to high	120 pieces per hour per fixture cavity	Medium and high	Setup time 1 h or less
Induction welding	Low to moderate	Low	900 pieces per hour, manually loaded	High	Setup time 1 h or less
Hot-gas welding	Very low	Low (holding fixture only)	0.3 to 1.5 m (12 to 60 in) of weld seam per minute	Very low	Manual operation

Table 18.48 Economic guide to a few welding processes

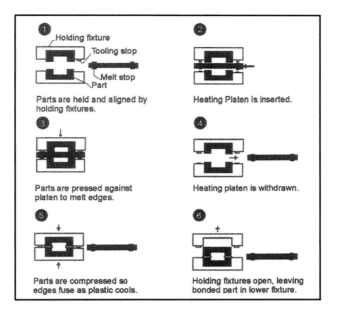

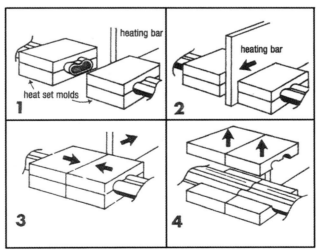

Figure 18.81 Example of hot-plate welding.

Weld	Material	Total Heating Time seconds	Plate Temperature °C	Heating and Welding Pressure MPa	Plate Removal Time seconds	Tensile Strength MPa
Weld 1	polypropylene	200	205	0.1	2	20.8
Weld 2	polypropylene	200	205	0.1	50	20.4
Weld 3	polypropylene	200	205	0.1	70	10.7
Parent Material	polypropylene					22.0

Table 18.49 Tensile strength of hot-plate welding PP copolymerized with ethylene pipe

Material					Falling Dart Impact For Weld and Bulk Samples			Weld Tensile Results		
Supplier	Grade	Melt Index	Density	Welding Temp.	Average weld impact energy	Average bulk impact energy	Weld factor (weld/bulk)	% failures during testing	Average stress @ failure	Average strain @ failure
		g/10 min	g/cc	°C	N-m [in-lb]	N-m [in-lb]		%	MPa [psi]	%
Quantum	LF6030	3.0	0.952	170				0		
Quantum	LF6040	5.5	0.962	160	24.6 [224]	33.0 [300]	0.75	50	26.1 [3792]	5.9
Quantum	LS5060	8.0	0.948	150	10.1 [92]	30.9 [281]	0.33	40	19.8 [2880]	10.2
Quantum	LS6060	10.0	0.962	140	12.1 [110]	33.0 [300]	0.37	100	17.1 [2480]	1.8
Quantum	LS6901	10.5	0.952	140	10.6 [96]	30.7 [279]	0.34	100	21.9 [3184]	5.5
Phillips	HMN4550	5.0	0.945	160	27.3 [248]	35.1 [319]	0.78	80	19.1 2768]	6.2
Phillips	HMN6060	7.5	0.963	150				100	24.5 [3552	5.4
Phillips	HMN6060-01	4.0	0.963	170				40	24.6 [3568]	5.4
Mobil	HMA045	7.0	0.952	150	20.7 [188]	33.0 [300]	0.62	40	18.2 [2640]	16.0
Mobil	NRA235	5.0	0.937	170	23.1 [210]	38.1 [346]	0.61	20	14.3 [2080]	13.6
Mobil	HRA034	3.5	0.940	200				0		
Dow	12350N	12.0	0.950	140	9.9 [90]	37.2 [338]	0.27	100	19.3 [2800]	6.8
Exxon	LL6407	7.0	0.935	150	14.0 [127]	34.4 [313]	0.40	20	14.1 [2048]	9.7

Table 18.50 Impact and tensile strength of hot-plate welding high-density polyethylene (HDPE)

Weld	Total Heating Time (seconds)	Plate Temperature (°C)	Heating and Welding Pressure (MPa)	Plate Removal Time (s)	Tensile Strength (MPa)
Weld 1	200	205	0.1	2	20.8
Weld 2	200	205	0.1	50	20.4
Weld 3	200	205	0.1	70	10.7
Parent Material					22.0

Table 18.51 Tensile strength of different hot-plate welds of PP copolymerized with ethylene pipe

ABS Sample	Exposure Type	Aging Temp.	Aging Time	Ultimate Tensile Strength		Elongation at Break	
		°C	days	MPa	% Retained	mm	% Retained
Hot Plate Weld	thermal air aging	120	3	19.7	61	0.9	42
Parent	thermal air aging	120	3	46.1	107	2.8	78
Hot Plate Weld	thermal air aging	120	7	20.5	63	0.9	41
Parent	thermal air aging	120	7	47.2	109	3.1	85
Hot Plate Weld	thermal air aging	120	14	12.8	40	0.7	33
Parent	thermal air aging	120	14	39.1	91	2.5	67
Hot Plate Weld	boiling water	100	3	9.3	29	0.6	29
Parent	boiling water	100	3	41.4	96	6.5	181
Hot Plate Weld	boiling water	100	7	12.6	38	0.7	31
Parent	boiling water	100	7	40.5	94	5.4	150

Table 18.52 Tensile strength of hot-plate welding ABS

In this process the surfaces to be joined are maintained under pressure against the hot plate, then the plate is withdrawn, and finally the matching surfaces are pressed together for joining. To avoid excessive lateral flow of melt out of the joint, some machines are equipped with rigid stops that bring the pressure automatically to zero once the part length is reduced to a fixed amount. This type of machine presents some advantages over the pressure-controlled machines and is particularly suitable for parts with small cross-sectional areas. Various types of tests have been used to evaluate the strength of the welds, including tensile, impact, and long-term creep-rupture tests. Microscopical examination has also been used to characterize hot-plate welds. However, the evaluation of the weld quality is still uncertain.

Welding Low-density Polyethylene

This review refers to low-density PE (LDPE) thin films that are produced by blown-film extrusion. In practice, the flat produced during this process is slit to thin film or used in tubular fashion to produce products such as bags. In either case, it is common to thermally weld these films during the fabrication process. The properties of the products produced depend on the strength of the film as extruded as well as the strength of the thermal weld used to join pieces of film together (249).

One important property in LDPE film applications is cold service temperature. As the service temperature for LDPE is lowered, the material and film undergo a ductile-to-brittle transition. Film properties on the brittle side of this transition exhibit significantly lower deformation to failure than is exhibited in ductile film. This lower ultimate deformation lowers impact properties and other properties related to ultimate deformation and the drawing process, such as toughness.

Thermal welding affects the low-temperature utility of these thin films by shifting the ductile-to-brittle transition temperature for the film and weld system to higher temperatures than the film without welds. Therefore, as the cold service temperature increases, the ductile-to-brittle transition temperature is encountered closer to ambient temperature, thus reducing the low-temperature performance of the welded film. It has been suggested that this shift in low-temperature utility arises from a geometric concentration of stress at the weld-film interface.

Mechanical property changes in a well-characterized area are termed the *heat-affected zone* (HAZ). The HAZ is defined as unmelted but changed material directly adjacent to a thermal weld. The reports of a HAZ are limited to the joining of larger parts. These larger parts were welded by methods such as butt-plate and hot-plate welding.

Larger parts cool more slowly and retain the weld energy longer than a thin film. Therefore, the development of a HAZ in large part is a function of thermal transport over much longer heating times. Thin film welding times are much shorter, typically 2 to 3 seconds versus minutes for large parts, and therefore the presence of a HAZ may not be a significant factor in the performance of thin film welds. A HAZ analogous to those observed in thick parts has never been reported in the case of thin film welds.

The problem is (1) to determine if there exists a HAZ present in thermally welded LDPE thin film, (2) to determine whether this HAZ has different properties than the original blown film, and (3) to determine if the shift in low-temperature ductile-brittle failures are caused by geometric stress concentration or the presence of a HAZ weaker than either the original blown film or the thermal weld.

Tests were performed on weld samples such as that shown in Figure 18.82. Evaluations were made based on mechanical testing and optical microscopy (chapter 22).

The optical microphotographs review provided strong visual evidence for the existence of a HAZ in these welded LDPE thin films. The measurements conducted observed HAZ scaled directly to the increase in heat input to the weld as measured by welding temperature. This observation is consistent with known models for thermal conductivity and literature reports of HAZ behavior.

Mechanical data provides strong evidence of affected mechanical properties in the welded LDPE thin film. In particular, the low-temperature performance of the welded film system is compromised by a shift in the temperature, where ductile-to-brittle transition occurs in the mechanical

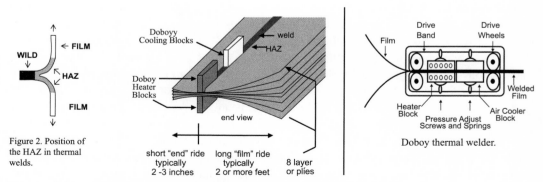

Figure 18.82 Film-welded, 8-ply arrangement using a Doboy thermal welder.

properties. The data demonstrates the weld region would be expected to fail long before the surrounding parent film.

A correlation was established between the increase in brittle failures and the width of the HAZ as observed by optical microscopy. This observation demonstrated the dependence of the HAZ width and the rate of brittle failures with welding temperature. If the geometry of the samples caused the failures, there would be no change in brittle failures with weld temperature. The increase in both HAZ width and brittle-failure rate with increasing temperature suggests that the increasing size of the HAZ is the key factor in the increased brittle failures seen in the welded LDPE thin film system.

Thermoband welding

This is a variation of the hot-plate welding method. A metallic tape acting as an electrical-resistance element is adhered to the material to be welded. Low voltage is applied to heat the material to its softening point so that the weld forms.

Hot-gas welding

In hot-gas or hot-air welding, a heated gas is used to heat TP parts and a filler rod to the melting or glass-transition temperature (chapter 1). The rod and the parts then soften and fuse, forming a high-strength bond upon cooling (Fig. 18.83 to 18.86). Hot-gas welding is commonly used for the fabrication and the repair of TP components and for lap welding of thin sheets or membranes. High bond strengths, up to 90% of the bulk material, can be achieved.

Infrared welding

In infrared welding, the joining surfaces of TP parts are heated to their melting temperature with infrared radiation at wavelengths ranging from 1 to 15 μm. When melting begins, parts are brought together under pressure, forming a weld upon cooling. Infrared welding is a noncontact welding method; Part surfaces are not in direct contact with the heat source but are at distances of up to about 20 mm. The high temperatures that can be obtained at short heating times makes this method especially suitable for temperature-resistant materials.

Infrared welding is similar to hot-tool welding. Joint surfaces of the parts to be welded are held at a specified distance from the heat source—a radiant heater or lamp—and are heated to the melt temperature of the plastic. When a molten layer of a desired thickness is obtained, the radiant source is removed, and the parts are brought together under pressure. The elapsed time from the removal of the heat source to the contact of the parts is the changeover phase. Pressure is applied to achieve intimate contact between the parts, and molecular diffusion across the joint interface determines joint strength. Pressure is applied until the joint cools and solidifies.

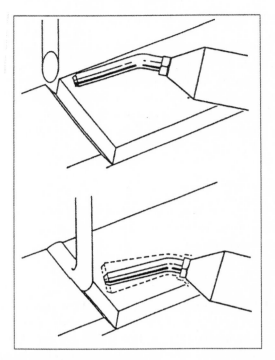

Figure 18.83 Example of a manual hot-gas welding.

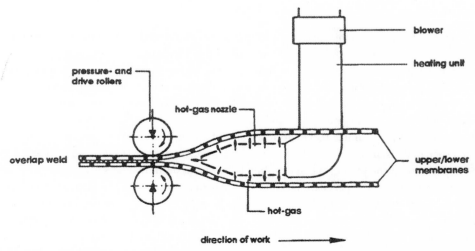

Figure 18.84 Example of an automatic hot-gas welder; hot gas blown between sheets, which melt and flow together.

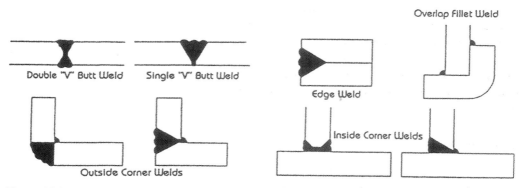

Figure 18.85 Example of design joints for hot-gas welding.

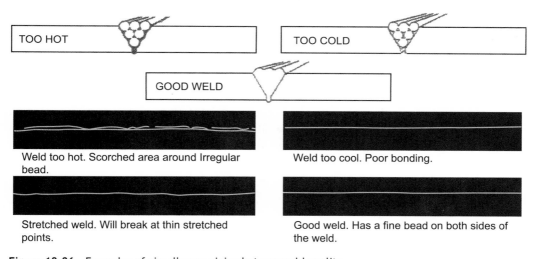

Figure 18.86 Examples of visually examining hot-gas weld quality.

VIBRATION WELDING

Vibration welding uses heat generated by friction at the interface of two TPs to produce melting in the interface area. The molten materials flow together under pressure and bond, forming a weld upon cooling. Vibration welding can be accomplished in a short time (8 to 15 seconds cycle time) and is applicable to a variety of plastic parts with planar or slightly curved surfaces. There are two types of vibration welding: linear, in which friction is generated by a linear back-and-forth motion, and orbital, in which the upper part to be joined is vibrated using circular motion in all directions.

Linear vibration welding is commonly used for flat parts. However, orbital-vibration welding makes the welding of irregularly shaped plastic parts possible.

In linear-vibration welding, the surfaces to be joined are rubbed together in an oscillating, linear motion under pressure applied at a 90° angle to the vibration. This process's parameters are the amplitude and frequency of this motion (weld amplitude and weld frequency), weld pressure, and weld time, which affect the strength of the resulting weld (Figs. 18.87 and 18.88 and Tables 18.53 to 18.60).

SPIN WELDING

Spin welding is a frictional process in which TP parts with rotationally symmetrical joining surfaces are rubbed together under pressure using unidirectional circular motion. The heat generated melts the plastic in the joining zone, forming a weld upon cooling. It is a fast, reliable process that requires only minimal, basic equipment but that can be completely automated. It is generally used for small cylindrical or spherical components (Figs. 18.89 and 18.90).

ULTRASONIC WELDING

Ultrasonic welding uses ultrasonic energy at high frequencies (20 to 40 kHz), that are beyond the range of human hearing, to produce low amplitude (1 to 25 μm, 0.001 to 0.025 mm, 0.000039 to 0.00098 in) mechanical vibrations. The vibrations generate heat at the joint interface of the parts being welded, resulting in the melt of the TP materials and weld formation after cooling (581, 582).

Ultrasonic welding is the fastest known welding process, with weld times of less than one second. In addition to welding, ultrasonic energy is commonly used for processes such as inserting metal into plastic parts or reforming TP parts to mechanically fasten dissimilar components.

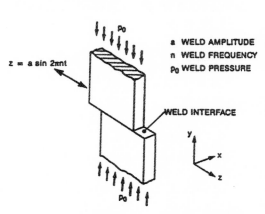

Figure 18.87 Example of linear-vibration welding.

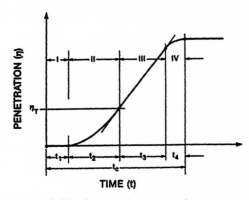

Figure 18.88 Penetration-versus-time curve showing the four phases of vibration welding.

Material Family	PC							
Tensile Strength[2], MPa (ksi)	68 (9.9)							
Elongation @ Break[2], %	6							
Specimen Thickness, mm (in.)	6.3 (0.25)	6.3 (0.25)	6.3 (0.25)	6.3 (0.25)	6.3 (0.25)	3.2 (0.125)	6.3 (0.25)	6.3 (0.25)
Mating Material								
Material Family[1]	ABS	M-PPO	M-PPO/PA	PC	PBT	PC/ABS	PC/PBT	PEI
Tensile Strength[2], MPa (ksi)	44 (6.4)	45.5 (6.6)	58 (8.5)	68 (9.9)	65 (9.5)	60 (8.7)	50 (7.3)	119 (17.3)
Elongation @ Break[2], %	2.2	2.5	>18	6	3.5	4.5		6
Specimen Thickness, mm (in.)	6.3 (0.25)	6.3 (0.25)	6.3 (0.25)	6.3 (0.25)	6.3 (0.25)	3.2 (0.125)	6.3 (0.25)	6.3 (0.25)
Process Parameters								
Process Type	vibration welding							
Weld Frequency	120 Hz							
Welded Joint Properties								
Weld Factor (weld strength/ weaker virgin material strength)	0.83	0.24	0.29	1.0	1.0	0.7	1.0	0.95
Elongation @ Break[2], % (nominal)	1.7	0.4	0.75	6	1.7	1.8	4.9	2.75

[1] ABS - acrylonitrile-butadiene-stryrene copolymer; M-PPO - modified polyphenylene oxide; M-PPO/PA - modified polyphenylene oxide/ polyamide alloy; PC - polycarbonate; PBT - polybutylene terephthalate polyester; PC/ABS - polycarbonate/ ABS alloy; PC/PBT - polycarbonate/ PBT alloy; PEI - polyetherimide
[2] strain rate of $10^{-2} s^{-1}$

Table 18.53 Properties of vibration welds of PC to itself and other plastics

Material Family	PC/ABS		
Tensile Strength[2], MPa (ksi)	60 (8.7)		
Elongation @ Break[2], %	4.5		
Specimen Thickness, mm (in.)	3.2 (0.125)	3.2 (0.125)	6.3 (0.25)
Mating Material			
Material Family[1]	ABS	PC	PC/ABS
Tensile Strength[2], MPa (ksi)	44 (6.4)	68 (9.9)	60 (8.7)
Elongation @ Break[2], %	1.8	6	4.5
Specimen Thickness, mm (in.)	3.2 (0.125)	3.2 (0.125)	6.3 (0.25)
Process Parameters			
Process Type	vibration welding		
Weld Frequency	120 Hz		
Welded Joint Properties			
Weld Factor (weld strength/ weaker virgin material strength)	0.85	0.7	0.85
Elongation @ Break[2], % (nominal)	1.8	1.8	2.3

[1] ABS - acrylonitrile-butadiene-stryrene copolymer; PC - polycarbonate; PC/ABS - polycarbonate/ ABS alloy
[2] strain rate of $10^{-2} s^{-1}$

Table 18.54 Properties of vibration welds of PC/ABS to itself and other plastics

Material Family	PC/PBT	
Tensile Strength[2], MPa (ksi)	50 (7.3)	
Specimen Thickness, mm (in.)	6.3 (0.25)	
Mating Material		
Material Family[1]	PC	PC/PBT
Tensile Strength[2], MPa (ksi)	68 (9.9)	50 (7.3)
Elongation @ Break[2], %	6	
Specimen Thickness, mm (in.)	3.2 (0.125)	6.3 (0.25)
Process Parameters		
Process Type	vibration welding	
Weld Frequency	120 Hz	
Welded Joint Properties		
Weld Factor (weld strength/ weaker virgin material strength)	1.0	1.0
Elongation @ Break[2], % (nominal)	4.9	>15

[1] PC - polycarbonate; PC/PBT - polycarbonate/ PBT alloy
[2] strain rate of $10^{-2} s^{-1}$

Table 18.55 Properties of vibration welds of PC/polybutylene terephthalate (PBT) to itself and to PC

Material Family	ABS					
Tensile Strength[2], MPa (ksi)	44 (6.4)					
Elongation @ Break[2], %	2.2					
Specimen Thickness, mm (in.)	6.3 (0.25)	6.3 (0.25)	6.3 (0.25)	6.3 (0.25)	3.2 (0.125)	6.3 (0.25)
Mating Material						
Material Family[1]	ABS	M-PPO	PC	PBT	PC/ABS	PEI
Tensile Strength[2], MPa (ksi)	44 (6.4)	45.5 (6.6)	68 (9.9)	65 (9.5)	60 (8.7)	119 (17.3)
Elongation @ Break[2], %	2.2	2.5	6	3.5	4.5	6
Specimen Thickness, mm (in.)	6.3 (0.25)	6.3 (0.25)	6.3 (0.25)	6.3 (0.25)	3.2 (0.125)	6.3 (0.25)
Process Parameters						
Process Type	vibration welding					
Weld Frequency	120 Hz					
Note						
Welded Joint Properties						
Weld Factor (weld strength/ weaker virgin material strength)	0.9	0.76	0.83	0.8	0.85	0.65
Elongation @ Break[2], % (nominal)	2.1	1.45	1.7	1.6	1.8	1.14

[1] ABS - acrylonitrile-butadiene-stryrene copolymer; ASA - Acrylonitrile-Stryrene-Acrylate Copolymer; M-PPO - modified polyphenylene oxide; M-PPO/PA - modified polyphenylene oxide/ polyamide alloy; PC - polycarbonate; PBT - polybutylene terephthalate polyester; PC/ABS - polycarbonate/ ABS alloy; PC/PBT - polycarbonate/ PBT alloy; PEI - polyetherimide
[2] strain rate of $10^{-2} s^{-1}$

Table 18.56 Properties of vibration welds of ABS to itself and other plastics

Material Family	ASA
Tensile Strength[2], MPa (ksi)	32.5 (4.7)
Elongation @ Break[2], %	2.9
Specimen Thickness, mm (in.)	6.3 (0.25)
Mating Material	
Material Family[1]	ASA
Tensile Strength[2], MPa (ksi)	32.5 (4.7)
Elongation @ Break[2], %	2.9
Specimen Thickness, mm (in.)	6.3 (0.25)
Process Parameters	
Process Type	vibration welding
Weld Frequency	120 Hz
Welded Joint Properties	
Weld Factor (weld strength/ weaker virgin material strength)	0.46
Elongation @ Break[2], % (nominal)	0.9

[1] ASA - Acrylonitrile-Stryrene-Acrylate Copolymer
[2] strain rate of $10^{-2} s^{-1}$

Table 18.57 Properties of vibration welds of acrylonitrile-styrene-acrylate (ASA) to itself

Material Family	M-PPO/PA		
Tensile Strength[2], MPa (ksi)	58 (8.5)		
Elongation @ Break[2], %	>18		
Specimen Thickness, mm (in.)	6.3 (0.25)		
Mating Material			
Material Family[1]	M-PPO/PA	M-PPO	PC
Tensile Strength[2], MPa (ksi)	58 (8.5)	45.5 (6.6)	68 (9.9)
Elongation @ Break[2], %	>18	2.5	6
Specimen Thickness, mm (in.)	6.3 (0.25)	6.3 (0.25)	6.3 (0.25)
Process Parameters			
Process Type	vibration welding		
Weld Frequency	120 Hz		
Welded Joint Properties			
Weld Factor (weld strength/ weaker virgin material strength)	1.0	0.22	0.29
Elongation @ Break[2], % (nominal)	>10	0.35	0.75

[1] M-PPO - modified polyphenylene oxide; M-PPO/PA - modified polyphenylene oxide/polyamide alloy; PC - polycarbonate;
[2] strain rate of $10^{-2} s^{-1}$

Table 18.58 Properties of vibration welds of PS-modified PPE/PA to itself and other plastics

Material Family	M-PPO				
Tensile Strength[2], MPa (ksi)	45.5 (6.6)				
Elongation @ Break[2], %	2.5				
Specimen Thickness, mm (in.)	6.3 (0.25)				
Mating Material					
Material Family[1]	ABS	M-PPO	M-PPO/PA	PC	PEI
Tensile Strength[2], MPa (ksi)	44 (6.4)	45.5 (6.6)	58 (8.5)	68 (9.9)	119 (17.3)
Elongation @ Break[2], %	2.2	2.5	>18	6	6
Specimen Thickness, mm (in.)	6.3 (0.25)	6.3 (0.25)	6.3 (0.25)	6.3 (0.25)	6.3 (0.25)
Process Parameters					
Process Type	vibration welding				
Weld Frequency	120 Hz				
Welded Joint Properties					
Weld Factor (weld strength/ weaker virgin material strength)	0.76	1.0	0.22	0.24	0
Elongation @ Break[2], % (nominal)	1.45	2.4	0.35	0.4	

[1] ABS - acrylonitrile-butadiene-stryrene copolymer; M-PPO - modified polyphenylene oxide; M-PPO/PA - modified polyphenylene oxide/ polyamide alloy; PC - polycarbonate; PEI - polyetherimide
[2] strain rate of $10^{-2} s^{-1}$

Table 18.59 Properties of vibration welds of modified polypropylene oxide (PPO) to itself and other plastics

Material Family	PBT			
Tensile Strength[2], MPa (ksi)	65 (9.5)			
Elongation @ Break[2], %	3.5			
Specimen Thickness, mm (in.)	6.3 (0.25)	6.3 (0.25)	6.3 (0.25)	6.3 (0.25)
Mating Material				
Material Family[1]	ABS	PC	PBT	PEI
Tensile Strength[2], MPa (ksi)	44 (6.4)	68 (9.9)	65 (9.5)	119 (17.3)
Elongation @ Break[2], %	2.2	6	3.5	6
Specimen Thickness, mm (in.)	6.3 (0.25)	6.3 (0.25)	6.3 (0.25)	6.3 (0.25)
Process Parameters				
Process Type	vibration welding			
Weld Frequency	120 Hz			
Welded Joint Properties				
Weld Factor (weld strength/ weaker virgin material strength)	0.8	1.0	0.96	0.95
Elongation @ Break[2], % (nominal)	1.6	1.7	3.5	4.1

[1] ABS - acrylonitrile-butadiene-stryrene copolymer; PC - polycarbonate; PBT - polybutylene terephthalate polyester; PEI - polyetherimide
[2] strain rate of $10^{-2} s^{-1}$

Table 18.60 Properties of vibration welds of PBT to itself and other plastics

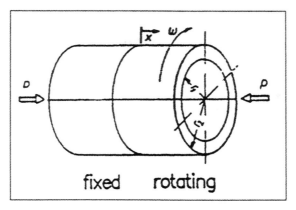

Figure 18.89 Spin welding, where one part does not move and the other part rotates.

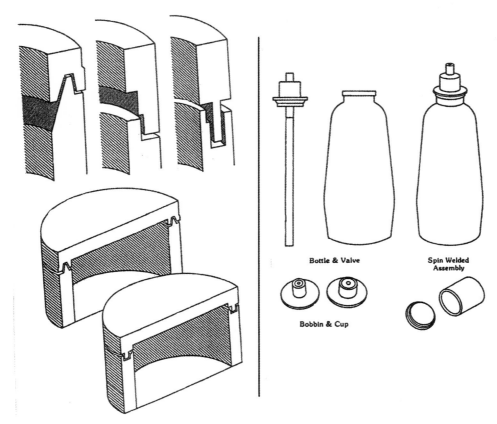

Figure 18.90 Example of a joint used in spin welding.

Components for ultrasonic welding are a generator or power supply, a converter or transducer, a booster, and a horn or sonotrode (Fig. 18.91). The generator converts low-voltage electricity at 50 to 60 Hz and 120 or 240 V to high-frequency (20 to 40 kHz) and high-voltage (13 kV) electrical energy. The electric current enters the converter, which contains piezoelectric ceramic crystals that expand and contract when excited by electrical energy. Electrical energy is converted into mechanical energy, and the converter expands and contracts at the frequency of the crystals.

Frequencies can range from 15 to 70 kHz; however, the most common frequencies used in ultrasonic welding are 20 or 40 kHz. The amplitude or peak-to-peak amplitude is the distance the converter moves back and forth during mechanical vibrations. Typical values are 20 μm (0.0008 in) for a 20 kHz converter and 9 μm (0.00035 in) for a 40 kHz converter.

The booster increases or decreases the amplitude of the mechanical vibrations of the converter, depending on the amplitude desired for welding, and conveys the vibrational energy to the horn or sonotrode. The horn, made of titanium or aluminum, can further increase the amplitude of the mechanical vibrations. It contacts one of the parts during welding and transmits vibratory energy to the part. For optimal energy transmission, the end of the horn that contacts the part is designed to mate with the part's geometry. Fixtures hold the parts in place and apply pressure during welding. An assembly stand is present in some welders in order to prevent movement or flexing of the weld-

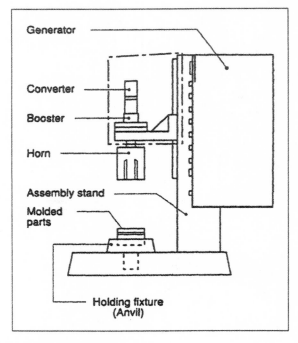

Figure 18.91 Components of an ultrasonic welder.

ing stack during welding. Ultrasonic welding begins after parts have been loaded and a particular force (trigger force) or distance has been reached by the horn.

Figure 18.92 shows the stages of ultrasonic welding. In Phase 1, the horn is placed in contact with the part (Fig. 18.93), pressure is applied, and vibratory motion is started. Heat generation due to friction melts the energy director, which then flows into the joint interface. The weld displacement begins to increase as the distance between the parts decreases. In Phase 2, the melting rate increases, resulting in increased weld displacement, and the part surfaces meet. Steady-state melting occurs in Phase 3, as a constant melt-layer thickness is maintained in the weld. In Phase 4, the holding phase, vibrations cease. Maximum displacement is reached, and intermolecular diffusion occurs as the weld cools and solidifies (Tables 18.61 to 18.63).

INDUCTION WELDING

Induction welding uses induction heating from a high-radio-frequency alternating current to magnetically excite ferromagnetic particles embedded in a TP or adhesive matrix at the joint interface of the two parts being welded. The heat released is used to melt and fuse TPs and heat hot-melt adhesives. It is a reliable and rapid technique, ranging from fractions of a second for small parts to 30 to 60 seconds for parts with long (30 cm; 157 in) weld areas, and results in structural, hermetic, or high-pressure welds (Figs. 18.94 and 18.95 and Table 18.64).

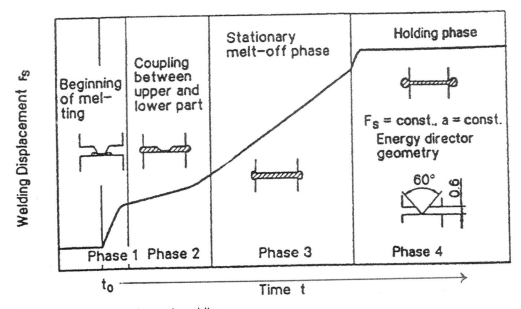

Figure 18.92 Stages in ultrasonic welding.

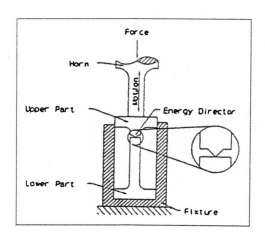 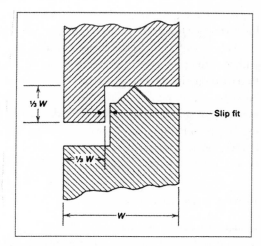

Figure 18.93 Examples of plastic mating joints to be ultrasonically welded.

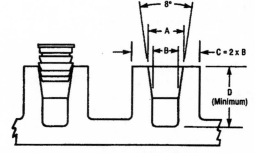

Thread Size	Insert OD "	Length "	Lead-in Dia A "	Diameter B "	Minimum Depth D "
4-40	.171	1/4	.160	.136	5/16
6-32	.217	5/16	.210	.177	3/8
8-32	.250	3/8	.240	.200	7/16
10-32	.295	7/16	.280	.235	1/2
1/4-20	.375	1/2	.355	.325	5/8

These are typical values. Since ultrasonic insets vary in size and design, contact manufacturer of inserts for their recommendations.

Table 18.61 Example of a boss-hole design for the use of ultrasonically installed inserts using styrene maleic anhydride copolymer

Mode	Welding Parameter	Energy Director (degrees)	
		60	90
constant energy mode	energy set value (Ws)	260	260
	delay timer (seconds)	0.3	0.1
	hold timer (seconds)	0.25	0.25
	hold pressure (kPa)	62	62
time based mode	weld timer (seconds)	0.5	0.5
	delay timer (seconds)	0.2	0.3
	hold timer (seconds)	0.65	0.65
	hold pressure (kPa)	104	104

Table 18.62 Optimum ultrasonic welding conditions for impact-modified PET-PC blend

#	Potential (KPa)		15	14	13	12	11	10	9	8	7	6	5	4	3	2	1
1.	5030	High Gloss High Impact ABS	EEE	SSS	GGE	GFG	FFF	XXX	EEG	EEG	FGF	EEE	FGG	SSS	EEE	EEE	SSS
2.	4650	Low Gloss, Medium Impact ABS	EEE	SSS	GEG	GGE	GGG	XXX	GEE	GGG	FGF	EEE	GFG	GGG	EGG	EEE	
3.	6205	Impact Modified RTPU	GGG	GFF	NW	NW	NW	NW	NW	NW	NW	GEG	FXX	GGG	GEG		
4.	8960	Clear, Low Impact RTPU	GGG	FFF	GGG	NW	NW	NW	XXX	FXX	FFF	GEE	FXX	GGG			
5.	10275	High Heat RTPU	WFF	FFF	FFG	FFF	FFF	XXX	FFF	FFF	WXX	GGF	FFF				
6.	4620	Medical Grade 75 Shore D TPU	FFF	FFF	NW	NW	NW	NW	XXX	FXX	FXX	EEE					
7.	3450	Medical Grade 55 Shore D TPU	FFF	FXX	NW	NW	NW	NW	NW	NW	XXX						
8.	8135	Medical Grade 15 MFR PC	FGF	GFG	GGF	FFF	FFF	FFF	SEE	SSS							
9.	8410	Gamma Stable 15 MFR PC	GGG	GGG	GFF	FFF	WFF	FFF	EEE								
10.	5410	Medium Molecular Weight GPPS	FFF	GFG	WWF	EEE	GGG	GGG									
11.	6790	High Molecular Weight GPPS	FFF	GFG	FFW	EEE	GGG										
12.	3965	High Impact Polystyrene HIPS	FFF	EGE	FFF	EEE											
13.	6825	Polycarbonate / ABS Blend	GGG	GGG	GGG												
14.	7860	Low Acrylonitrile SAN	EFG	EGE													
15.	9030	High Acrylonitrile SAN	GGG														

NW - No Bonding Occurred in the Weld Area
W - Weak Bond (<5% of Potential)
F - Fair Bond (6 - 15% of Potential)
G - Good Bond (16 - 30% of Potential)
E - Excellent (31 - 50% of Potential)
S - Superior (over 50% of Potential)
X - No Data

Potential - Refers to the force necessary to break a homogeneous piece, with no weld, of the same structure. If different resins are welded together, it refers to the potential of the weaker resin.

Table 18.63 Weld strength of ultrasonic bonds of medical plastics; three letters in each box represent bonds subjected to no sterilization, ethylene-oxide sterilization, and gamma-radiation sterilization, respectively

Induction welding is a type of electromagnetic welding that uses electromagnetic energy at frequencies of 0.1 kHz to 10 MHz to heat materials. Induction welding is frequently referred to as electromagnetic welding.

Heat is generated in induction welding from interaction of the magnetic field with the ferromagnetic material and from current induced in the metal. The high-frequency alternating current moves through a copper coil, generating a rapidly reversing magnetic field. Ferromagnetic materials

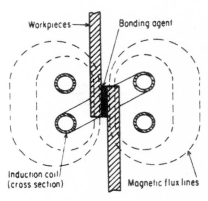

Figure 18.94 Example of induction heat produced during induction welding.

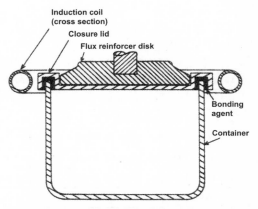

Figure 18.95 Example of induction welding a lid to a container.

	ABS	ABS/PC alloy	Acetal	Acrylic	Cellulosics	Modified PPO	Polybutylene	Polycarbonate	Polyethylene	UHMW-PE	Polypropylene	Polystyrene	Polysulfone	Polyurethane	PVC	SAN	TPE (Hytrel)	TPE (Kraton)	TPE (Ren)	TP polyester	Paper
ABS	●	●		●				●				●				●					●
ABS/PC alloy	●	●		●				●					●								●
Acetal			●																		●
Acrylic	●	●		●				●					●								●
Cellulosics					●																●
Modified PPO						●						●									●
Polybutylene							●														●
Polycarbonate	●	●		●				●													●
Polyethylene									●									●			●
UHMW-PE										●											●
Polypropylene											●								●		●
Polystyrene						●						●									●
Polysulfone		●						●					●								●
Polyurethane	●													●							●
PVC															●						●
SAN	●			●								●				●					●
TPE (Hytrel)																	●				●
TPE (Kraton)									●									●			●
TPE (Ren)											●								●		●
TP polyester																				●	●
Paper	●	●	●	●	●	●	●	●	●	●	●	●	●	●	●	●	●	●	●	●	●

a—Bullets indicate compatible combinations.

Table 18.64 Guide to bonding plastic to plastic via induction welding

align with the magnetic field; alignment changes as magnetic field direction changes. Atoms do not return to the initial alignment existing before the magnetic field changed direction but to a slightly different alignment, resulting in heat losses within the metal, which are transferred to the plastic matrix through conduction. Additional heat generation results from voltage within metallic, ferromagnetic material induced by alternating current.

Radio-frequency welding

Radio-frequency welding, also called high-frequency welding or sealing, heat sealing, and dielectric welding or sealing, uses high-frequency (13 MHz to 100 MHz) electromagnetic energy to generate heat in polar materials, resulting in melting and weld formation after cooling. A high-intensity radio signal is used to impart increased molecular motion in two similar or dissimilar plastics. This causes a temperature rise, resulting in melting and increased plastic-chain mobility. Ultimately, the plastic chains of the two materials penetrate their interface and become entangled, forming a weld. Tables 18.65 to 18.67 provide data where the breaking-strength-per-unit cross-sectional area of each weld was calculated and then divided by the tensile strength of the weaker material. This number (multiplied by 100) gave the weld strength expressed as a percentage of the highest possible value or "potential."

Microwave welding

Microwave welding uses high-frequency electromagnetic radiation to heat a susceptor material located at the joint interface. The generated heat melts TP materials at the joint interface, producing a weld upon cooling. Microwave welding is a type of electromagnetic welding and uses frequencies of 2 to 10 GHz.

Heat generation occurs through absorption of microwave energy by susceptor materials that contain polar groups as part of their molecular structure or that are electrically conductive. In an applied electric field, polar groups align in the field direction. In a microwave, the magnitude and direction of the electric field changes rapidly; polar molecules develop strong oscillations as they continually align with the field, generating heat through friction.

Polyaniline (PAN) doped with an aqueous acid such as HCl is used as a susceptor in microwave welding. Doping with dilute aqueous acid introduces polar groups into the molecular structure and makes the material electrically conductive by providing free-moving electrons. The amount of heat produced during welding is dependent on the conductivity. If the material is not conductive enough, mobility of free charges is low and very little heating occurs. If conductivity is too high, microwave energy is reflected, not absorbed, so that no heating occurs. PAN is an "A-B" type of polymer with conductivity ranging from that of an insulator to 400 S/cm, depending on the doping material used. For welding TPs, acid-doped PAN powder and a TP material is compression molded into a gasket that is placed at the joint interface. Bulk material in the parts being joined is not affected by microwave heating unless the molecular structure includes polar groups.

Material	Joining Material	RF welded without cutting through samples	Samples purposely cut through during RF welding		Change in weld strength after exposure to 5-5.5 Mrads of gamma radiation
			no aging	aged 48 hours at 60°C	
clear flexbile PVC 65A	TPE alloy	no bond	no bond		
clear flexbile PVC 65A	styrenic TPE	no bond	no bond		
clear flexbile PVC 65A	aromatic polyester polyurethane	excellent (31-50% potential)	good (16-30% potential)	fair (6-15% potential)	
clear flexbile PVC 65A	filled radiopaque PVC 75A	superior (>50% potential)	fair (6-15% potential)	good (16-30% potential)	
clear flexbile PVC 65A	clear rigid PVC 80D	fair (6-15% potential)	excellent (31-50% potential)	excellent (31-50% potential)	0% potential
clear flexbile PVC 65A	clear flexible PVC 80A	superior (>50% potential)	good (16-30% potential)	good (16-30% potential)	
clear flexbile PVC 65A	clear flexible PVC 65A	good (16-30% potential)	good (16-30% potential)	good (16-30% potential)	+1% potential
clear flexible PVC 80A	TPE alloy	no bond	no bond		
clear flexible PVC 80A	styrenic TPE	no bond	no bond		
clear flexible PVC 80A	aromatic polyester polyurethane	superior (>50% potential)	good (16-30% potential)	fair (6-15% potential)	
clear flexible PVC 80A	filled radiopaque PVC 75A	excellent (31-50% potential)	good (16-30% potential)	good (16-30% potential)	
clear flexible PVC 80A	clear rigid PVC 80D	excellent (31-50% potential)	good (16-30% potential)	superior (>50% potential)	+3% potential
clear flexible PVC 80A	clear flexible PVC 80A	superior (>50% potential)	good (16-30% potential)	good (16-30% potential)	+5% potential
filled radiopaque PVC 75A	TPE alloy	no bond	no bond		
filled radiopaque PVC 75A	styrenic TPE	no bond	no bond		
filled radiopaque PVC 75A	aromatic polyester polyurethane	excellent (31-50% potential)	good (16-30% potential)	fair (6-15% potential)	
filled radiopaque PVC 75A	filled radiopaque PVC 75A	excellent (31-50% potential)	good (16-30% potential)	good (16-30% potential)	

Table 18.65 Properties of radio-frequency welding of flexible PVC to itself and other plastics

Material	Joining Material	RF welded without cutting through samples	Samples purposely cut through during RF welding		Change in weld strength after exposure to 5-5.5 Mrads of gamma radiation
			no aging	aged 48 hours at 60°C	
clear rigid PVC 80D	TPE alloy	no bond	no bond		
clear rigid PVC 80D	styrenic TPE	no bond	no bond		
clear rigid PVC 80D	aromatic polyester polyurethane	fair (6-15% potential)	excellent (31-50% potential)	superior (>50% potential)	
clear rigid PVC 80D	filled radiopaque PVC 75A	fair (6-15% potential)	superior (>50% potential)	excellent (31-50% potential)	
clear rigid PVC 80D	clear rigid PVC 80D	weak (<5% potential)	fair (6-15% potential)	fair (6-15% potential)	-5% potential
clear rigid PVC 80D	clear flexible PVC 80A	excellent (31-50% potential)	good (16-30% potential)	superior (>50% potential)	
clear rigid PVC 80D	clear flexible PVC 65A	fair (6-15% potential)	excellent (31-50% potential)	excellent (31-50% potential)	

Table 18.66 Properties of radio-frequency welding of rigid PVC to itself and other plastics

Material	Joining Material	RF welded without cutting through samples	Samples purposely cut through during RF welding	
			no aging	aged 48 hours at 60°C
aromatic polyester polyurethane	TPE alloy	no bond	no bond	
aromatic polyester polyurethane	styrenic TPE	no bond	no bond	
aromatic polyester polyurethane	aromatic polyester polyurethane	fair (6-15% potential)	excellent (31-50% potential)	fair (6-15% potential)
aromatic polyester polyurethane	filled radiopaque PVC 75A	excellent (31-50% potential)	good (16-30% potential)	fair (6-15% potential)
aromatic polyester polyurethane	clear rigid PVC 80D	fair (6-15% potential)	excellent (31-50% potential)	superior (>50% potential)
aromatic polyester polyurethane	clear flexible PVC 80A	superior (>50% potential)	good (16-30% potential)	fair (6-15% potential)
aromatic polyester polyurethane	clear flexible PVC 65A	excellent (31-50% potential)	good (16-30% potential)	fair (6-15% potential)

Table 18.67 Properties of radio-frequency welding of aromatic polyester PUR to itself and other plastics

Resistance welding

In resistance welding, also called resistance-implant welding, a current is applied to a conductive heating element or implant placed at the joint interface of the parts being welded. The implant is heated through Joule heating, and the surrounding plastic melts and flows together, forming a weld. Joint strengths of TPs are higher with this welding method than those obtained with adhesive bonding.

Heating elements can be carbon-fiber prepregs (chapter 15), woven graphite fabric, or stainless steel foil or mesh. Stainless steel heating elements can be used alone in welding TPs; for welding TSs or metals, they are impregnated with a TP or sandwiched between two TP layers, such as glass/PEEK coated with thin layers of PEEK film.

Carbon-fiber prepregs are reinforced plastics usually composed of unidirectional carbon fibers in a TP matrix. Stainless steel heating elements introduce foreign material into the joint but are used to minimize any potential for galvanic corrosion and to reduce fiber motion, which has a detrimental effect on joint strength. The heating element comprises the innermost portion of the weld stack (Fig. 18.96) used in resistance welding. The heating element is sandwiched between the parts to be joined. Insulators on the outermost ends of the weld stack complete the assembly. The weld stack can be autoclaved for consolidation.

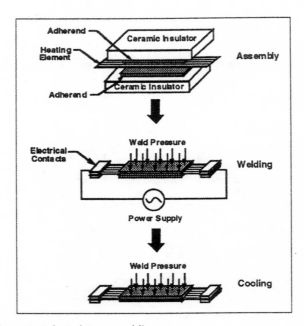

Figure 18.96 The three steps in resistance welding.

Extrusion welding

Extrusion welding is a reliable process developed from hot-gas welding, in which a TP filler identical to the material being welded is extruded into a groove in the preheated weld area (chapter 5). The filler material fills the groove and forms a weld after cooling. Extrusion welding is usually performed manually, although it can be automated (Fig. 18.97).

Laser welding

Laser welding is a process in which a high-intensity laser beam is used to increase the temperature at the joint interface of TP materials to or above the melt temperature. The molten plastics cool and solidify, forming a weld (583).

Carbon dioxide (CO_2) and Nd-YAG (neodymium ions in a medium of yttrium aluminum garnet) lasers are predominantly used in industrial applications. CO_2 lasers emit radiation at wavelengths of 9.2 to 10.8 µm, with the strongest emission at 10.6 µm, and range in power from 30 W to 40 kW. The laser beam is transmitted through air, reflected from mirrors, and focused using zinc-selenide (ZnSe) lenses. In the Nd-YAG laser, flash lamps excite Nd ions in a solid crystal rod, resulting in radiation with the strongest emission at a wavelength of 1.06 µm.

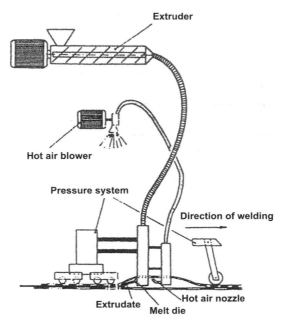

Figure 18.97 Example of an extrusion-welding system, where the hot air melts the plastic to be welded prior to the extruded melt flows into the area.

The short-wavelength beam is transmitted through a fiber-optic beam delivery system, and power ranges from 30 W to 2 kW. Lasers can generate radiation continuously (continuous wave), or light may be emitted in short bursts of microsecond or millisecond duration (pulsed); pulsed lasers are useful when overheating of the material is a problem. CO_2 lasers generally operate in a pulsed mode, while either pulsed or continuous-wave modes can be used with Nd-YAG lasers.

In laser welding, the parts being joined are clamped onto a moving table. Pressure can be applied throughout the process or may only be applied after heating is terminated. In the presence of a shielding gas, the high-intensity laser beam travels at a high speed across the weld interface of the parts being joined, cutting the weld interface and homogeneously heating the weld area. The diameter of the dot-like laser beam corresponds approximately to the wall thickness of the parts being joined. The beam causes heating that is localized near the joint interface that can rapidly result in melting, degradation, and vaporization of the polymer in the weld interface (Tables 18.68 and 18.69).

MACHINING

OVERVIEW

Although most plastic parts are usually fabricated into their final shapes, there are parts that require secondary machining from parts that need supplemental operations (cutting extruded shape, cutting molded gates, cutting thermoformed scrap, etc.) to taking stock plastics to produce parts via

Material			Laser Conditions			Tensile Properties	
Type	Thickness, mm	Joint type	Type	Power, W	Speed, m/min	% of parent	Failure mode
PE	0.1	lap	CO_2	100	16.5	>100	parent
PE	0.1	lap	CO_2	200	36		
PE	0.1	lap	CO_2	300	50		
PE	0.1	cut/seal	CO_2	100	5.7	94	weld
PE	0.5	lap	Nd:YAG	80	0.1	68	weld

Table 18.68 Properties of laser-welded PE joints

Material			Laser Conditions			Tensile Properties	
Type	Thickness, mm	Joint type	Type	Power, W	Speed, m/min	% of parent	Failure mode
PP	0.2	lap	CO_2	100	51	98	parent
PP	0.2	lap	Nd:YAG	80	0.2	70	weld
PP	2.0	butt	Nd:YAG	80	0.1	30	weld

Table 18.69 Properties of laser-welded PP joints

machining (443, 584–592). Different machining operations are involved: milling, drilling, cutting, finishing, and so on (Table 18.70). Different reasons exist for machining, such as the following:

1. Dimensions of fabricated parts may not be sufficiently accurate. Extreme accuracy, particularly when using certain processes or machines with limited capabilities, can be expensive to achieve.
2. Fabricated parts can be relatively expensive in small-scale production. It may also be desirable to make parts by machining when the production is not large enough to justify the investment in fabricating equipment.
3. Supplementary machining procedures may be required in finishing operations, as shown in Table 18.71. Examples of machining operations on parts are shown in Table 18.72.

There is a variety of machining characteristics among the many types of plastics. TPs are relatively resilient compared to metals, and they require special cutting procedures (Fig. 18.98). Even a TP's cutting characteristic will change depending on the fillers and reinforcements used (chapter 1).

Machining method	Purpose of machining operation
Cutting	
with a single-point tool	Turning, planing, shaping
with a multiple-point tool	Milling, drilling, reaming, threading, engraving
Cutting off	
with a saw	Hack sawing, band sawing, circular sawing
by the aid of abrasives	Bonded abrasives: abrasive cutting off, diamond cutting off
	Loose abrasives: blasting, ultrasonic cutting off
shearing	Shearing, nibbling
by the aid of heat	Friction cutting off, electrical heated wire cutting off
Finishing	
by the aid of abrasives	Bonded abrasives: grinding, abrasive belt grinding
	Loose abrasives: barreling, blasting, buffing

Table 18.70 Examples of machining operations

Types of parts	Kinds of machining methods used
Bearing, roller	Turning, milling, drilling, shaping
Button	Turning, drilling
Cam	Turning, copy turning
Dial and scale	Engraving, sand blasting
Gear	Turning, milling, gear shaving, broaching
Liner and brake lining	Cutting off, shaping, planing, milling
Pipe and rod	Cutting off, turning, threading
Plate (ceiling, panel)	Cutting off, drilling, tapping
Tape (mainly for PTFE)	Peeling

Table 18.71 Examples of finishing operations

Processing method	Purpose of machining operation	Types of machining used
Compression, transfer, injection and blow molding	Degating, deflashing, polishing	Cutting off, buffing, tumbling, filing, sanding
Extrusion	Cut lengths of extrudate	Cutting off
Laminating	Cut sheets to size, deflashing edges	Cutting off
Vacuum forming	Polish cut edges, trim parts to size	Cutting off, sanding, filing

Table 18.72 Examples of supplementary machining operations

Elastic recovery occurs in plastics both during and after machining; provisions must be made in the tool geometry for sufficient clearance to allow for it. This is due to the expansion of any compressed material that takes place because elastic recovery causes increased friction between the recovered cut surface and the cutting surface of the tool. In addition to generating heat, this abrasion affects tool wear. Elastic recovery also explains why, without proper precautions, drilled or tapped holes in many plastics often are tapered or become smaller than the diameter of the drills that were used to make them (particularly if TPs are unfilled or not reinforced).

The heat of conductivity of plastics is very low. Essentially all the cutting heat generated will be absorbed by the cutting tool. The small amount of heat conducted into the plastic cannot be transferred to the core of the shape, so it causes the heat of the surface area to increase significantly. If this heat is kept to a minimum, no further action is required. Otherwise, heat removed by a coolant is used to ensure a proper cut. For many commodity TPs, the softening, deformation, and degradation heats are relatively low. Gumming, discoloration, poor tolerance control, and poor finish could occur if frictional heat is generated and allowed to buildup. Engineered TPs (PA, polytetrafluoroethylene [PTFE], etc.; chapter 1) have relatively high melting or softening points. Thus they have less of a tendency to become gummed, melted, burned, or crazed in machining than do plastics with lower melting points.

TS plastics machining is slightly different than TPs because there is not any great melting distortion from a fast cutting speed. Higher cutting speeds improve machined finishes. However, the added frictional heat can reduce tool life, and the surface of the plastic to be machined can also be distorted in appearance by burning unless precautionary steps are taken, such as spraying a coolant directly on the cutting tool and plastic. Another major difference in machining TSs is the type of chips that are removed by the cutting tool. Almost all of these are in a powderlike form that can be readily removed with the aid of a vacuum hose. This is almost a must in moderate- and high-production runs due to contamination of the air in the vicinity of the cutting machine. In some machining instances, the dust condition created can be hazardous enough to compel the operator to wear a filtering air mask.

All plastics can be properly machined or cut when a few simple rules are observed: (1) use only sharp tools, (2) provide adequate chip clearance, (3) support the work properly, and (4) provide adequate cooling. Dull tools do not cut properly, which results in a poor surface finish. Because they

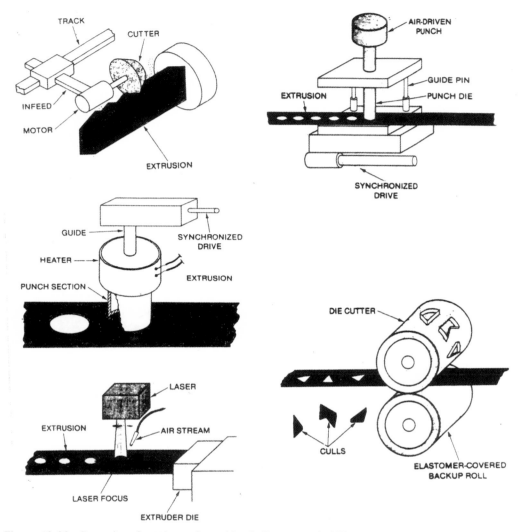

Figure 18.98 Examples of cutting and punching in-line, extruded TPs.

require greater pressure for cutting, there is unnecessary deflection of the workpiece and excessive frictional heat buildup. Well-sharpened tools scrape properly, leaving a good finish on the work, and remain serviceable for a reasonable length of time.

Adequate chip clearance prevents clogging of the tool and interference with cutting edges. When there is a choice of tools, the one with the greatest provision for chip clearance should be used. For example, drills with wide flute areas and saw blades with deep gullets are preferred. The work should be properly supported to prevent springing away from the tool under cutting pressure. Excessive deflection of the work by the cutting tool causes chattering or uneven cutting.

CHARACTERISTICS

The main consideration characteristics for plastics materials (chapter 2) and tooling are summarized in the following list:

1. *Low thermal conductivity.* Heat conductivity of plastics is substantially less than that of metal. Essentially all the heat generated by cutting friction between the plastic and the cutting tool will be absorbed by the metal cutting tool. The small amount of heat conducted into the plastic cannot be transferred to the core of the part, and the temperature of the surface layer will rise significantly. This heat must be kept minimal or be removed by a coolant to insure a good job.
2. *Low modulus of elasticity.* TPs are relatively resilient (soft) compared to metals; therefore the forces involved in holding and cutting must be adjusted accordingly and the material properly supported to prevent distortion. This characteristic will vary to some extent in different families of plastics. For example, vinyl, PE, and TFE fluoroplastics have relatively low resistance to deformation, whereas polystyrene is harder and usually more brittle than either of the other three.
3. *Plastic memory.* Elastic recovery occurs in plastic materials both during and after machining. Thus provisions must be made in the tool geometry for sufficient clearance to provide relief. This is because the expansion of compressed material, due to elastic recovery, causes increased friction between the recovered cut surface and the relief surface of the tool. In addition to generating heat, this abrasion causes tool wear. Elastic recovery that occurs after machining also explains why, if proper precautions are not used, drilled or tapped holes in plastics often are tapered. They can also become smaller than the diameter of the drills used to make them and also why turned diameters often become larger than the dimensions measured just after cutting.
4. *Coefficient of thermal expansion.* The coefficients of thermal expansion of the usual TPs are greater, roughly by ten times, than those of metals. Expansion of the plastics caused by the heat generated during machining increases friction and consequently the amount of heat produced. Adequate cutting-tool clearances are necessary to avoid rubbing.
5. *Softening point.* The softening, deformation, and degradation temperatures of TPs are relatively low. Gumming, discoloration, poor tolerance control, and poor finish can occur if frictional heat is generated and allowed to build up. TPs have relatively high melting or softening points, as is the case with nylon or fluoroplastics. They are less prone to gumming, melting, and crazing during machining than are plastics with lower melting points. Heat buildup becomes more critical in plastics with lower melting points.

Machining and Cutting Operations

There are production runs of plastic parts that are machined from raw plastic stock. With many configurations, high production sometimes can be more economically achieved by machining than by precision molding. Other considerations, such as delivery date or the availability and cost of tooling and production equipment, may dictate machining instead of molding. Each type of plastic has unique properties and therefore has different machining characteristics from those of the metallic materials as summarized in Tables 18.73 to 18.85. These tables provide guides. The machine-tool industry can be divided into the two broad groups of cutting machines (milling, turning, grinding, etc.) and forming machines (bending, shearing [Fig. 18.99], etc.).

In many extrusion-film and coating operations, the slitting and winding must be dealt with as one operation. Figure 18.100 highlights some of the factors that are involved, including (a) the effect of torque on web tension in center winding; (b) the top-riding roll riding on gauge band; (c) an example of razor blade slitting in air and in grooved roll; and (d) a guide to cutting.

Some tips for tool geometry when machining plastics are the following:

1. Reduce frictional drag and temperature by honing or polishing surfaces of the tool where it comes in contact with the plastic.
2. Geometries of tools should be such that they generate continuous-type chips. In general, large rake angles will serve this purpose because of the force directions resulting from these rake angles. Care must be exercised so that rake angles will not be so large that brittle fracture of the workpiece results and chips become discontinuous.
3. The drill geometry for plastics should differ from that of metals in that wide, polished flutes combined with low helix angles should be used to help eliminate the packing of chips that causes overheating. Also important is that the normal 118° point angle is generally modified to 70° to 120° (Table 18.86).

Figure 18.99 Example of extrusion in-line shear cutter with sheets being stacked.

Material	Hardness	Condition	Depth of cut, in.	High-speed steel tool			Carbide tool			
				Speed, ft./min.	Feed, in./rev.	Tool mat'l	Speed Brazed, ft./min.	Indexable, ft./min.	Feed, in./rev.	Tool mat'l
Acrylic, acetal, polycarbonate, polysulfone, polystyrene	60 to 120 R_M	Cast, molded, or extruded	0.040 0.150 0.300 0.625	400 350 300 —	0.005 0.008 0.010 —	M2, M3 M2, M3 M2, M3 —	600 550 500 —	600 550 500 —	0.005 0.010 0.012 —	C-2 C-2 C-2 —
ABS, polyarylether, polypropylene, polyethylene, cellulose acetate	50 to 120 R_R	Cast, molded, or extruded	0.040 0.150 0.300 0.625	450 400 350 —	0.005 0.008 0.010 —	M2, M3 M2, M3 M2, M3 —	700 650 600 —	700 650 600 —	0.005 0.010 0.010 —	C-2 C-2 C-2 —
Fluoroplastics: TFE, CTFE	74 to 95 R_R	Molded or extruded	0.040 0.150 0.300 0.625	400 350 300 —	0.005 0.008 0.010 —	M2, M3 M2, M3 M2, M3 —	600 550 500 —	600 550 500 —	0.005 0.010 0.012 —	C-2 C-2 C-2 —
Polyamides (nylons): 6, 6/6, 6/12, 11, 12	78 to 120 R_R	Molded or extruded	0.040 0.150 0.300 0.625	500 450 400 —	0.005 0.010 0.012 —	M2, M3 M2, M3 M2, M3 —	800 700 650 —	800 700 650 —	0.005 0.010 0.012 —	C-2 C-2 C-2 —
Polyamides (nylons): 35% glass-reinforced, types 6, 6/6	78 to 120 R_R	Filled and molded	0.040 0.150 0.300 0.625	— — — —	— — — —	— — — —	600 550 500 —	600 550 500 —	0.005 0.008 0.010 —	C-2 C-2 C-2 —
Polyamides (nylons): 35% glass-reinforced, types 6/10, 6/12	40 to 50 R_E	Filled and molded	0.040 0.150 0.300 0.625	— — — —	— — — —	— — — —	500 450 400 —	500 450 400 —	0.005 0.008 0.010 —	C-2 C-2 C-2 —
Epoxy, melamine, phenolic	100 to 128 R_M	Cast or molded	0.040 0.150 0.300 0.625	500 450 400 —	0.005 0.010 0.015 —	M2, M3 M2, M3 M2, M3 —	800 700 650 —	800 700 650 —	0.005 0.010 0.015 —	C-2 C-2 C-2 —
Furan, polybutadiene	40 to 100 R_R	Cast	0.040 0.150 0.300 0.625	250 200 175 —	0.005 0.010 0.015 —	T15, M42[b] T15, M42[b] T15, M42[b] —	450 400 350 —	450 400 350 —	0.005 0.008 0.010 —	C-2 C-2 C-2 —
Silicone	15 to 65 Shore A	Cast or molded	0.040 0.150 0.300 0.625	200 175 150 —	0.005 0.010 0.015 —	T15, M42 T15, M42 T15, M42 —	450 400 350 —	450 400 350 —	0.005 0.008 0.010 —	C-2 C-2 C-2 —
Silicone, glass-filled	80 to 90 R_M	Filled and molded	0.040 0.150 0.300 0.625	— — — —	— — — —	— — — —	400 350 300 —	400 350 300 —	0.005 0.008 0.010 —	C-2 C-2 C-2 —
Polyimide	40 to 50 R_E	Molded or extruded	0.040 0.150 0.300 0.625	500 450 400 —	0.005 0.010 0.015 —	M2, M3 M2, M3 M2, M3 —	800 700 700 —	800 700 700 —	0.005 0.010 0.015 —	C-2 C-2 C-2 —
Polyimide, glass-filled	109 to 115 R_M	Filled and molded	0.040 0.150 0.300 0.625	— — — —	— — — —	— — — —	500 450 400 —	500 450 400 —	0.005 0.010 0.012 —	C-2 C-2 C-2 —
Polyurethane	65 to 95 Shore A	Cast	0.040 0.150 0.300 0.625	250 200 175 —	0.005 0.010 0.015 —	T15, M42 T15, M42 T15, M42 —	450 400 350 —	450 400 350 —	0.005 0.008 0.010 —	C-2 C-2 C-2 —
Polyurethane	55 to 75 Shore D	Cast	0.040 0.150 0.300 0.625	300 250 200 —	0.005 0.010 0.015 —	M2, M3 M2, M3 M2, M3 —	500 450 400 —	500 450 400 —	0.005 0.010 0.015 —	C-2 C-2 C-2 —
Allyl (DAP)	95 to 100 R_M	Cast	0.040 0.150 0.300 0.625	400 350 300 —	0.005 0.008 0.012 —	M2, M3 M2, M3 M2, M3 —	600 550 500 —	600 550 500 —	0.005 0.010 0.015 —	C-2 C-2 C-2 —

Table 18.73 Guide to single-point box-tool machining (chapter 17 reviews tool materials)

Material	Hardness	Condition	Speed, ft./min.	Feed, in./rev. Cutoff tool width, in.			Form tool width, in.					Tool mat'l
				0.062	0.125	0.205	0.500	0.750	1.00	1.50	2.00	
Acrylic, acetal, polycarbonate, polysulfone, polystyrene	60 to 120 R_M	Cast, molded, or extruded	250 400	0.002 0.002	0.003 0.003	0.004 0.004	0.008 0.008	0.006 0.006	0.005 0.005	0.004 0.004	0.003 0.003	M2, M3 C-2
ABS, polyarylether, polypropylene, polyethylene, cellulose acetate	50 to 120 R_R	Cast, molded, or extruded	300 500	0.002 0.002	0.003 0.003	0.004 0.004	0.008 0.008	0.006 0.006	0.005 0.005	0.004 0.004	0.003 0.003	M2, M3 C-2
Fluoroplastics: TFE, CTFE	74 to 95 R_R	Molded or extruded	250 400	0.002 0.002	0.003 0.003	0.004 0.004	0.008 0.008	0.006 0.006	0.005 0.005	0.004 0.004	0.003 0.003	M2, M3 C-2
Polyamides (nylons): 6, 6/6, 6/12, 11, 12	78 to 120 R_R	Molded or extruded	350 600	0.003 0.003	0.005 0.005	0.006 0.006	0.012 0.012	0.010 0.010	0.008 0.008	0.006 0.006	0.004 0.004	M2, M3 C-2
Polyamides (nylons): 35% glass-reinforced, types 6, 6/6	78 to 120 R_R	Filled and molded	— 450	— 0.002	— 0.003	— 0.004	— 0.008	— 0.006	— 0.005	— 0.004	— 0.003	— C-2
Polyamides (nylons): 35% glass-reinforced, types 6/10, 6/12	40 to 50 R_E	Filled and molded	— 350	— 0.002	— 0.003	— 0.004	— 0.008	— 0.006	— 0.005	— 0.004	— 0.003	— C-2
Epoxy, melamine, phenolic	100 to 128 R_M	Cast or molded	350 600	0.003 0.003	0.005 0.005	0.006 0.006	0.012 0.012	0.010 0.010	0.008 0.008	0.006 0.006	0.004 0.004	M2, M3 C-2
Furan, polybutadiene	40 to 100 R_R	Cast	175 325	0.001 0.001	0.002 0.002	0.003 0.003	0.007 0.007	0.005 0.005	0.004 0.004	0.003 0.003	0.002 0.002	M2, M3 C-2
Silicone	15 to 65 Shore A	Cast or molded	150 300	0.001 0.001	0.002 0.002	0.003 0.003	0.007 0.007	0.005 0.005	0.004 0.004	0.003 0.003	0.002 0.002	T15, M42 C-2
Silicone, glass-filled	80 to 90 R_M	Filled and molded	— 250	— 0.002	— 0.003	— 0.004	— 0.007	— 0.005	— 0.004	— 0.003	— 0.002	— C-2
Polyimide	40 to 50 R_E	Molded or extruded	350 600	0.003 0.003	0.005 0.005	0.006 0.006	0.012 0.012	0.010 0.010	0.008 0.008	0.006 0.006	0.004 0.004	M2, M3 C-2
Polyimide, glass-filled	109 to 115 R_M	Filled and molded	— 375	— 0.002	— 0.003	— 0.004	— 0.007	— 0.005	— 0.004	— 0.003	— 0.002	— C-2
Polyurethane	65 to 95 Shore A	Cast	150 300	0.001 0.001	0.002 0.002	0.003 0.003	0.007 0.007	0.005 0.005	0.004 0.004	0.003 0.003	0.002 0.002	T15, M42 C-2
Polyurethane	55 to 75 Shore D	Cast	175 350	0.002 0.002	0.003 0.003	0.004 0.004	0.008 0.008	0.006 0.006	0.005 0.005	0.004 0.004	0.003 0.003	M2, M3 C-2
Allyl (DAP)	95 to 100 R_M	Cast	250 400	0.002 0.002	0.003 0.003	0.004 0.004	0.007 0.007	0.005 0.005	0.004 0.004	0.003 0.003	0.002 0.002	M2, M3 C-2

Table 18.74 Guide to turning, cutoff, and form-tool machining

Machining may be required on fabricated plastics—blocks of plastics, simple to complicated molded or cast shapes, extruded or protruded profiles, sheet, film, rod, or tube—to turn them into finished products. A problem when machining plastics, particularly TPs, is the heat built up by friction. As the plastic and cutting tools begin to heat up, the plastic can distort or melt. This can result in a poor surface finish, tearing, localized melting, welding of stacked products, and jamming of cutters.

It is important to prevent the product and cutting tool from heating up to the point at which significant softening or melting takes place. There are cutting tools specifically designed to cut plastic that eliminate or reduce the heating problem. Some plastic materials can machine much easier

Material	Hardness	Condition	Speed, ft./min.	Feed, in./rev. Nominal hole diameter								Tool mat'l
				1/16	1/8	1/4	1/2	3/4	1	1½	2	
Acrylic, acetal, polycarbonate, polysulfone, polystyrene	60 to 120 R_M	Cast, molded, or extruded	100 200	0.001 —	— 0.002	— 0.004	— 0.005	— 0.006	— 0.008	— 0.010	— 0.012	M10, M7, M1
ABS, polyarylether, polypropylene, polyethylene, cellulose acetate	50 to 120 R_R	Cast, molded, or extruded	150 250	0.001 —	— 0.002	— 0.004	— 0.005	— 0.006	— 0.008	— 0.010	— 0.012	M10, M7, M1
Fluoroplastics: TFE, CTFE	75 to 95 R_R	Molded or extruded	100 200	0.001 —	— 0.002	— 0.004	— 0.005	— 0.006	— 0.008	— 0.010	— 0.012	M10, M7, M1
Polyamides (nylons): 6, 6/6, 6/12, 11, 12	78 to 120 R_R	Molded or extruded	100 250	0.001 —	— 0.002	— 0.004	— 0.005	— 0.006	— 0.008	— 0.010	— 0.012	M10, M7, M1
Polyamides (nylons): 35% glass-reinforced types 6, 6/6	78 to 120 R_R	Filled and molded	75 150	0.001 —	— 0.002	— 0.004	— 0.005	— 0.006	— 0.008	— 0.010	— 0.012	T15, M42[b]
Polyamides (nylons): 35% glass-reinforced types 6/10, 6/12	40 to 50 R_E	Filled and molded	65 125	0.001 —	— 0.002	— 0.004	— 0.005	— 0.006	— 0.008	— 0.010	— 0.012	T15, M42[b]
Epoxy, melamine, phenolic	100 to 128 R_M	Cast or molded	100 200	0.001 —	— 0.002	— 0.003	— 0.004	— 0.005	— 0.006	— 0.008	— 0.010	M10, M7, M1
Furan, polybutadiene	40 to 100 R_R	Cast	100 200	0.001 —	— 0.002	— 0.003	— 0.004	— 0.005	— 0.006	— 0.008	— 0.010	M10, M7, M1
Silicone	15 to 65 Shore A	Cast or molded	65 125	0.001 —	— 0.002	— 0.003	— 0.004	— 0.006	— 0.008	— 0.008	— 0.010	M10, M7, M1
Silicone, glass-filled	80 to 90 R_M	Filled and molded	65 125	0.001 —	— 0.001	— 0.002	— 0.003	— 0.004	— 0.006	— 0.008	— 0.010	T15, M42[b]
Polyimide	40 to 50 R_E	Molded or extruded	100 200	0.001 —	— 0.002	— 0.004	— 0.005	— 0.006	— 0.008	— 0.010	— 0.012	M10, M7, M1
Polyimide, glass-filled	109 to 115 R_M	Filled and molded	65 125	0.001 —	— 0.001	— 0.002	— 0.003	— 0.004	— 0.006	— 0.008	— 0.010	T15, M42
Polyurethane	65 to 95 Shore A	Cast	100 200	0.001 —	— 0.002	— 0.003	— 0.004	— 0.005	— 0.006	— 0.008	— 0.010	M10, M7, M1
Polyurethane	55 to 75 Shore D	Cast	150 250	0.001 —	— 0.002	— 0.004	— 0.005	— 0.006	— 0.008	— 0.010	— 0.012	M10, M7, M1
Allyl (DAP)	95 to 100 R_M	Cast	100 200	0.001 —	— 0.002	— 0.004	— 0.005	— 0.006	— 0.008	— 0.010	— 0.012	M10, M7, M1

Table 18.75 Guide to drilling

and faster than others due to their physical and mechanical properties. Generally, a high melting point, inherent lubricity, and good hardness and rigidity are factors that improve machinability.

The fabricating methods divide into three broad categories: (1) the machining of solid shapes; (2) the cutting, sewing, and sealing of film and sheeting; and (3) the forming of films and sheets. The following list describes these methods in greater detail:

1. *Machining*. Using techniques quite common to the metal and the wood industries, plastic shapes can be turned into end products by such methods as grinding, turning on a lathe, sawing, reaming, milling, routing, drilling, and tapping.

Material	Hardness	Condition	Axial depth of cut, in.	Speed, ft./min.	Feed, in./tooth Width of slot, in.				Tool mat'l
					3/8	1/2	3/4	1-2	
Acrylic, acetal, polycarbonate, polysulfone, polystyrene	60 to 120 R_M	Cast, molded, or extruded	0.030 0.125 diam./2 diam./1	400 350 300 250	0.003 0.003 0.002 0.001	0.004 0.003 0.002 0.001	0.005 0.004 0.003 0.002	0.006 0.005 0.004 0.003	M2, M3, M7
ABS, polyarylether, polypropylene, polyethylene, cellulose acetate	50 to 120 R_R	Cast, molded, or extruded	0.030 0.125 diam./2 diam./1	500 450 400 350	0.003 0.003 0.002 0.001	0.004 0.003 0.002 0.001	0.005 0.004 0.003 0.002	0.006 0.005 0.004 0.003	M2, M3, M7
Fluoroplastics: TFE, CTFE	74 to 95 R_R	Molded or extruded	0.030 0.125 diam./2 diam./1	400 350 300 250	0.003 0.003 0.002 0.001	0.004 0.003 0.002 0.001	0.005 0.004 0.003 0.002	0.006 0.005 0.004 0.003	M2, M3, M7
Polyamides (nylons): 6, 6/6, 6/12, 11, 12	78 to 120 R_R	Molded or extruded	0.030 0.125 diam./2 diam./1	500 450 400 350	0.003 0.003 0.002 0.001	0.004 0.003 0.002 0.001	0.005 0.004 0.003 0.002	0.006 0.005 0.004 0.003	M2, M3, M7
Polyamides (nylons): 35% glass-reinforced, types 6, 6/6	78 to 120 R_R	Filled and molded	0.030 0.125 diam./2 diam./1	400 350 300 250	0.003 0.003 0.002 0.001	0.003 0.003 0.002 0.001	0.004 0.003 0.003 0.002	0.005 0.005 0.004 0.003	C-2 (carbide)
Polyamides (nylons): 35% glass-reinforced, types 6/10, 6/12	40 to 50 R_E	Filled and molded	0.030 0.125 diam./2 diam./1	350 300 250 200	0.003 0.003 0.002 0.001	0.003 0.003 0.002 0.001	0.004 0.003 0.003 0.002	0.005 0.005 0.004 0.003	C-2 (carbide)
Epoxy, melamine, phenolic	100 to 128 R_M	Cast or molded	0.030 0.125 diam./2 diam./1	500 450 400 350	0.003 0.003 0.002 0.001	0.004 0.004 0.003 0.002	0.008 0.008 0.006 0.004	0.010 0.010 0.008 0.006	M2, M3, M7
Furan, polybutadiene	40 to 100 R_R	Cast	0.030 0.125 diam./2 diam./1	300 275 250 200	0.003 0.003 0.002 0.001	0.004 0.004 0.003 0.002	0.005 0.005 0.004 0.003	0.006 0.006 0.005 0.004	T15, M42
Silicone	15 to 65 Shore A	Cast or molded	0.030 0.125 diam./2 diam./1	300 275 250 200	0.003 0.003 0.002 0.001	0.004 0.004 0.003 0.002	0.005 0.005 0.004 0.003	0.006 0.006 0.005 0.004	T15, M42
Silicone, glass-filled	80 to 90 R_M	Filled and molded	0.030 0.125 diam./2 diam./1	350 300 275 250	0.003 0.003 0.002 0.001	0.004 0.004 0.003 0.002	0.005 0.005 0.004 0.003	0.006 0.006 0.005 0.004	C-2 (carbide)
Polyimide	40 to 50 R_E	Molded or extruded	0.030 0.125 diam./2 diam./1	500 450 400 350	0.003 0.003 0.002 0.001	0.004 0.004 0.003 0.002	0.008 0.008 0.006 0.004	0.010 0.010 0.008 0.006	M2, M3, M7
Polyimide, glass-filled	109 to 115 R_M	Filled and molded	0.030 0.125 diam./2 diam./1	450 400 350 300	0.003 0.003 0.002 0.001	0.004 0.004 0.003 0.002	0.005 0.005 0.004 0.003	0.006 0.006 0.005 0.004	C-2 (carbide)
Polyurethane	65 to 95 Shore A	Cast	0.030 0.125 diam./2 diam./1	300 275 250 200	0.003 0.003 0.002 0.001	0.004 0.004 0.003 0.002	0.005 0.005 0.004 0.003	0.006 0.006 0.005 0.004	T15, M42
Polyurethane	55 to 75 Shore D	Cast	0.030 0.125 diam./2 diam./1	350 300 275 250	0.003 0.003 0.002 0.001	0.004 0.004 0.003 0.002	0.006 0.006 0.005 0.004	0.008 0.008 0.007 0.006	M2, M3, M7
Allyl (DAP)	95 to 100 R_M	Cast	0.030 0.125 diam./2 diam./1	350 300 275 250	0.003 0.003 0.002 0.001	0.004 0.004 0.003 0.002	0.006 0.006 0.005 0.004	0.008 0.008 0.006 0.005	M2, M3, M7

Table 18.76 Guide to end milling: Slotting machining

Material	Hardness	Condition	Radial depth of cut, in.	High-speed steel tool						Carbide tool					
				Speed ft./min.	Feed, in./tooth — Cutter diameter, in.				Tool mat'l	Speed ft./min.	Feed, in./tooth — Cutter diameter, in.				Tool mat'l
					3/8	1/2	3/4	1-2			3/8	1/2	3/4	1-2	
Acrylic, acetal, polycarbonate, polysulfone, polystyrene	60 to 120 R_M	Cast, molded, or extruded	0.020 0.060 diam./4 diam./2	500 475 450 400	0.004 0.004 0.005 0.003	0.006 0.006 0.008 0.005	0.010 0.010 0.012 0.008	0.015 0.015 0.018 0.012	M2, M3, M7	1300 1200 1000 800	0.004 0.004 0.005 0.003	0.006 0.006 0.008 0.005	0.010 0.010 0.012 0.008	0.015 0.015 0.018 0.012	C-2
ABS, polyarylether, polypropylene, polyethylene, cellulose acetate	50 to 120 R_R	Cast, molded, or extruded	0.020 0.060 diam./4 diam./2	650 550 400 300	0.004 0.004 0.005 0.003	0.006 0.006 0.008 0.005	0.010 0.010 0.012 0.008	0.015 0.015 0.018 0.012	M2, M3, M7	1400 1300 1200 1000	0.004 0.004 0.005 0.003	0.006 0.006 0.008 0.005	0.010 0.010 0.012 0.008	0.015 0.015 0.018 0.012	C-2
Fluoroplastics: TFE, CTFE	74 to 95 R_R	Molded or extruded	0.020 0.060 diam./4 diam./2	500 475 450 400	0.004 0.004 0.005 0.003	0.006 0.006 0.008 0.005	0.010 0.010 0.012 0.008	0.015 0.015 0.018 0.012	M2, M3, M7	1300 1200 1000 800	0.004 0.004 0.005 0.003	0.006 0.006 0.008 0.005	0.010 0.010 0.012 0.008	0.015 0.015 0.018 0.012	C-2
Polyamides (nylons): 6, 6/6, 6/12, 11, 12	78 to 120 R_R	Molded or extruded	0.020 0.060 diam./4 diam./2	650 550 400 300	0.004 0.004 0.005 0.003	0.006 0.006 0.008 0.005	0.010 0.010 0.012 0.008	0.015 0.015 0.018 0.012	M2, M3, M7	1400 1300 1200 1000	0.004 0.004 0.005 0.003	0.006 0.006 0.008 0.005	0.010 0.010 0.012 0.008	0.015 0.015 0.018 0.012	C-2
Polyamides (nylons), 35% glass-reinforced, types 6, 6/6	78 to 120 R_R	Filled and molded	0.020 0.060 diam./4 diam./2	— — — —	— — — —	— — — —	— — — —	— — — —	—	600 550 500 450	0.004 0.004 0.005 0.003	0.005 0.005 0.006 0.004	0.007 0.007 0.008 0.006	0.010 0.010 0.012 0.008	C-2
Polyamides (nylons), 35% glass-reinforced, types 6/10, 6/12	40 to 50 R_E	Filled and molded	0.020 0.060 diam./4 diam./2	— — — —	— — — —	— — — —	— — — —	— — — —	—	550 500 450 400	0.004 0.004 0.005 0.003	0.005 0.005 0.006 0.004	0.007 0.007 0.008 0.006	0.010 0.010 0.012 0.008	C-2
Epoxy, melamine, phenolic	100 to 128 R_M	Cast or molded	0.020 0.060 diam./4 diam./2	650 550 400 300	0.004 0.004 0.005 0.003	0.006 0.006 0.008 0.005	0.010 0.010 0.012 0.008	0.015 0.015 0.018 0.012	M2, M3, M7	1400 1300 1200 1000	0.004 0.004 0.005 0.003	0.006 0.006 0.008 0.005	0.010 0.010 0.012 0.010	0.015 0.015 0.018 0.012	C-2
Furan, polybutadiene	40 to 100 R_R	Cast	0.020 0.060 diam./4 diam./2	350 300 275 250	0.004 0.004 0.005 0.003	0.006 0.006 0.008 0.005	0.010 0.010 0.012 0.008	0.015 0.015 0.018 0.012	T15	800 750 600 550	0.004 0.004 0.005 0.003	0.006 0.006 0.008 0.005	0.010 0.010 0.012 0.010	0.015 0.015 0.018 0.012	C-2
Silicone	15 to 65 Shore A	Cast or molded	0.020 0.060 diam./4 diam./2	300 250 200 175	0.004 0.004 0.005 0.003	0.006 0.006 0.008 0.005	0.010 0.010 0.012 0.008	0.015 0.015 0.018 0.012	T15	600 550 500 450	0.004 0.004 0.005 0.003	0.006 0.006 0.008 0.005	0.010 0.010 0.012 0.010	0.015 0.015 0.018 0.012	C-2
Silicone, glass-filled	80 to 90 R_M	Filled and molded	0.020 0.060 diam./4 diam./2	— — — —	— — — —	— — — —	— — — —	— — — —	—	500 450 400 350	0.004 0.004 0.005 0.003	0.006 0.006 0.008 0.005	0.010 0.010 0.012 0.010	0.015 0.015 0.018 0.012	C-2
Polyimide	40 to 50 R_E	Molded or extruded	0.020 0.060 diam./4 diam./2	650 550 400 300	0.004 0.004 0.005 0.003	0.006 0.006 0.008 0.005	0.010 0.010 0.012 0.008	0.015 0.015 0.018 0.012	M2, M3, M7	1400 1300 1200 1000	0.004 0.004 0.005 0.003	0.006 0.006 0.008 0.005	0.010 0.010 0.012 0.010	0.015 0.015 0.018 0.012	C-2
Polyimide, glass-filled	109 to 115 R_M	Filled and molded	0.020 0.060 diam./4 diam./2	— — — —	— — — —	— — — —	— — — —	— — — —	—	600 550 500 450	0.004 0.004 0.005 0.003	0.005 0.005 0.006 0.004	0.007 0.007 0.008 0.006	0.010 0.010 0.012 0.008	C-2
Polyurethane	65 to 95 Shore A	Cast	0.020 0.060 diam./4 diam./2	350 300 275 250	0.004 0.004 0.005 0.003	0.006 0.006 0.008 0.005	0.010 0.010 0.012 0.008	0.015 0.015 0.018 0.012	M2, M3, M7	800 750 600 550	0.004 0.004 0.005 0.003	0.006 0.006 0.008 0.005	0.010 0.010 0.012 0.010	0.015 0.015 0.018 0.012	C-2
Polyurethane	55 to 75 Shore D	Cast	0.020 0.060 diam./4 diam./2	400 350 300 250	0.004 0.004 0.005 0.003	0.006 0.006 0.008 0.005	0.010 0.010 0.012 0.008	0.015 0.015 0.018 0.012	M2, M3, M7	900 850 800 750	0.004 0.004 0.005 0.003	0.006 0.006 0.008 0.005	0.010 0.010 0.012 0.010	0.015 0.015 0.018 0.012	C-2
Allyl (DAP)	95 to 100 R_M	Cast	0.020 0.060 diam./4 diam./2	400 350 300 250	0.004 0.004 0.005 0.003	0.006 0.006 0.008 0.005	0.010 0.010 0.012 0.008	0.015 0.015 0.018 0.012	M2, M3, M7	1000 900 850 750	0.004 0.004 0.005 0.003	0.006 0.006 0.008 0.005	0.010 0.010 0.012 0.008	0.015 0.015 0.018 0.012	C-2

Table 18.77 Guide to end milling: Peripheral machining

Material	Hardness	Condition	Depth of cut, in	High-speed steel tool			Carbide tool			
				Speed ft./min.	Feed per tooth, in.	Tool mat'l, AISI	Speed Brazed, ft./min.	Indexable, ft./min.	Feed per tooth, in.	Tool mat'l, grade
Acrylic, acetal, polycarbonate, polysulfone, polystyrene	60 to 120 R_M	Cast, molded, or extruded	0.040 0.150 0.300	300 260 220	0.004 0.005 0.007	M2, M3 M2, M3 M2, M3	475 450 400	475 450 400	0.004 0.005 0.006	C-2 C-2 C-2
ABS, polyarylether, polypropylene, polyethylene, cellulose acetate	50 to 120 R_R	Cast, molded, or extruded	0.040 0.150 0.300	340 300 260	0.004 0.005 0.007	M2, M3 M2, M3 M2, M3	550 525 475	550 525 475	0.004 0.005 0.006	C-2 C-2 C-2
Fluoroplastics: TFE, CTFE	74 to 95 R_R	Molded or extruded	0.040 0.150 0.300	300 260 220	0.004 0.005 0.007	M2, M3 M2, M3 M2, M3	475 450 400	475 450 400	0.003 0.004 0.005	C-2 C-2 C-2
Polyamides (nylons): 6, 6/6, 6/12, 11, 12	78 to 120 R_R	Molded or extruded	0.040 0.150 0.300	380 340 300	0.004 0.006 0.008	M2, M3 M2, M3 M2, M3	650 550 525	650 550 525	0.004 0.005 0.007	C-2 C-2 C-2
Polyamides (nylons): 35% glass-reinforced, types 6, 6/6	78 to 120 R_R	Filled and molded	0.040 0.150 0.300	— — —	— — —	— — —	475 450 400	475 450 400	0.003 0.004 0.005	C-2 C-2 C-2
Polyamides (nylons): 35% glass-reinforced, types 6/10, 6/12	40 to 50 R_E	Filled and molded	0.040 0.150 0.300	— — —	— — —	— — —	400 380 340	400 380 340	0.003 0.004 0.005	C-2 C-2 C-2
Epoxy, melamine, phenolic	100 to 128 R_M	Cast or molded	0.040 0.150 0.300	380 340 260	0.004 0.005 0.007	M2, M7 M2, M7 M2, M7	650 550 525	650 550 525	0.003 0.004 0.005	C-2 C-2 C-2
Furan, polybutadiene	40 to 140 R_R	Cast	0.040 0.150 0.300	200 150 115	0.004 0.005 0.007	T15, M42 T15, M42 T15, M42	360 320 280	360 320 280	0.003 0.004 0.005	C-2 C-2 C-2
Silicone	15 to 65 Shore A	Cast or molded	0.040 0.150 0.300	150 135 95	0.004 0.005 0.007	T15, M42 T15, M42 T15, M42	360 320 280	360 320 280	0.003 0.004 0.005	C-2 C-2 C-2
Silicone, glass-filled	80 to 90 R_M	Filled and molded	0.040 0.150 0.300	— — —	— — —	— — —	320 280 240	320 280 240	0.003 0.004 0.005	C-2 C-2 C-2
Polyimide	40 to 50 R_E	Molded or extruded	0.040 0.150 0.300	380 340 260	0.004 0.005 0.007	M2, M7 M2, M7 M2, M7	650 550 475	650 550 475	0.004 0.005 0.007	C-2 C-2 C-2
Polyimide, glass-filled	109 to 115 R_M	Filled and molded	0.040 0.150 0.300	— — —	— — —	— — —	390 360 320	390 360 320	0.004 0.005 0.007	C-2 C-2 C-2
Polyurethane	65 to 95 Shore A	Cast	0.040 0.150 0.300	200 150 115	0.004 0.005 0.007	T15, M42 T15, M42 T15, M42	360 320 280	360 320 280	0.003 0.004 0.005	C-2 C-2 C-2
Polyurethane	55 to 75 Shore D	Cast	0.040 0.150 0.300	220 200 150	0.004 0.005 0.007	M2, M7 M2, M7 M2, M7	390 360 320	390 360 320	0.003 0.004 0.005	C-2 C-2 C-2
Allyl (DAP)	95 to 100 R_M	Cast	0.040 0.150 0.300	300 260 220	0.004 0.005 0.007	M2, M3 M2, M3 M2, M3	475 450 400	475 450 400	0.004 0.005 0.007	C-2 C-2 C-2

Table 18.78 Guide to side and slot milling arbor-mounted cutter machining

Material	Hardness	Condition	Depth of cut, in.	High-speed steel tool			Carbide tool			
				Speed ft./min.	Feed per tooth, in.	Tool mat'l	Speed		Feed per tooth, in.	Tool mat'l
							Brazed, ft./min.	Indexable, ft./min.		
Acrylic, acetal, polycarbonate, polysulfone, polystyrene	60 to 120 R_M	Cast, molded, or extruded	0.040 0.150 0.300	400 350 300	0.005 0.008 0.010	M2, M3 M2, M3 M2, M3	650 600 550	650 600 550	0.005 0.007 0.009	C-2 C-2 C-2
ABS, polyarylether, polypropylene, polyethylene, cellulose acetate	50 to 120 R_R	Cast, molded, or extruded	0.040 0.150 0.300	450 400 350	0.005 0.008 0.010	M2, M3 M2, M3 M2, M3	750 700 650	750 700 650	0.004 0.007 0.009	C-2 C-2 C-2
Fluoroplastics: TFE, CTFE	74 to 95 R_R	Molded or extruded	0.040 0.150 0.300	400 350 300	0.005 0.008 0.010	M2, M3 M2, M3 M2, M3	650 600 550	650 600 550	0.004 0.007 0.009	C-2 C-2 C-2
Polyamides (nylons): 6, 6/6, 6/12, 11, 12	78 to 120 R_R	Molded or extruded	0.040 0.150 0.300	500 450 400	0.006 0.010 0.014	M2, M3 M2, M3 M2, M3	850 750 700	850 750 700	0.006 0.008 0.010	C-2 C-2 C-2
Polyamides (nylons): 35% glass-reinforced, types 6, 6/6	78 to 120 R_R	Filled and molded	0.040 0.150 0.300	— — —	— — —	— — —	650 600 550	650 600 550	0.004 0.006 0.008	C-2 C-2 C-2
Polyamides (nylons): 35% glass-reinforced, types 6/10, 6/12	40 to 50 R_E	Filled and molded	0.040 0.150 0.300	— — —	— — —	— — —	550 500 450	550 500 450	0.004 0.006 0.008	C-2 C-2 C-2
Epoxy, melamine, phenolic	100 to 128 R_M	Cast or molded	0.040 0.150 0.300	500 450 350	0.005 0.008 0.010	M2, M7 M2, M7 M2, M7	850 750 700	850 750 700	0.004 0.006 0.008	C-2 C-2 C-2
Furan, polybutadiene	40 to 100 R_R	Cast	0.040 0.150 0.300	250 200 150	0.005 0.008 0.010	T15, M42 T15, M42 T15, M42	75 425 375	475 425 375	0.004 0.006 0.008	C-2 C-2 C-2
Silicone	15 to 65 Shore A	Cast or molded	0.040 0.150 0.300	200 175 125	0.005 0.008 0.010	T15, M42 T15, M42 T15, M42	475 425 375	475 425 375	0.004 0.006 0.008	C-2 C-2 C-2
Silicone, glass-filled	80 to 90 R_M	Filled and molded	0.040 0.150 0.300	— — —	— — —	— — —	425 375 325	425 375 325	0.004 0.006 0.008	C-2 C-2 C-2
Polyimide	40 to 50 R_E	Molded or extruded	0.040 0.150 0.300	500 450 350	0.005 0.008 0.010	M2, M7 M2, M7 M2, M7	850 750 650	850 750 650	0.005 0.008 0.010	C-2 C-2 C-2
Polyimide, glass-filled	109 to 115 R_M	Filled and molded	0.040 0.150 0.300	— — —	— — —	— — —	525 475 425	525 475 425	0.005 0.008 0.010	C-2 C-2 C-2
Polyurethane	65 to 95 Shore A	Cast	0.040 0.150 0.300	250 200 150	0.005 0.008 0.010	T15, M42 T15, M42 T15, M42	475 425 375	475 425 375	0.004 0.006 0.008	C-2 C-2 C-2
Polyurethane	55 to 75 Shore D	Cast	0.040 0.150 0.300	300 250 200	0.005 0.008 0.010	M2, M7 M2, M7 M2, M7	525 475 425	525 475 425	0.004 0.006 0.008	C-2 C-2 C-2
Allyl (DAP)	95 to 100 R_M	Cast	0.040 0.150 0.300	400 350 300	0.005 0.008 0.010	M2, M3 M2, M3 M2, M3	650 600 550	650 600 550	0.006 0.008 0.010	C-2 C-2 C-2

Table 18.79 Guide to face-milling machining

Material	Hardness	Condition	Material thickness, in.	Tooth form	Teeth/in.	Band speed, ft./min.
Acrylic, acetal, polycarbonate, polysulfone, polystyrene	60 to 120 R_M	Cast, molded, or extruded	<½ ½-1 1-3 >3	P P B B	8-14 6-8 3 3	4000 3500 3000 2500
ABS, polyarylether, polypropylene, polyethylene, cellulose acetate	50 to 120 R_R	Cast, molded, or extruded	<½ ½-1 1-3 >3	P P B B	8-14 6-8 3 3	4500 4000 3500 3000
Fluoroplastics: TFE, CTFE	74 to 95 R_R	Molded or extruded	<½ ½-1 1-3 >3	P P B B	8-14 6-8 3 3	4000 3500 3000 2500
Polyamides (nylons): 6, 6/6, 6/12, 11, 12	78 to 120 R_R	Molded or extruded	<½ ½-1 1-3 >3	P P B B	8-14 6-8 3 3	5000 4300 3500 3000
Epoxy, melamine, phenolic	100 to 128 R_M	Cast or molded	<½ ½-1 1-3 >3	P P B B	8-14 6-8 3 3	5000 4300 3500 3000
Furan, polybutadiene	40 to 100 R_R	Cast	<½ ½-1 1-3 >3	P P B B	8-14 6-8 3 3	2500 2200 1800 1500
Silicone	15 to 65 Shore A	Cast or molded	<½ ½-1 1-3 >3	P P B B	8-14 6-8 3 3	2000 1800 1500 1200
Polyimide	40 to 50 R_E	Molded or extruded	<½ ½-1 1-3 >3	P P B B	8-14 6-8 3 3	5000 4300 3500 3000
Polyurethane	65 to 95 Shore A	Cast	<½ ½-1 1-3 >3	P P B B	8-14 6-8 3 3	2500 2200 1800 1500
Polyurethane	55 to 75 Shore D	Cast	<½ ½-1 1-3 >3	P P B B	8-14 6-8 3 3	3000 2700 2200 1800
Allyl (DAP)	95 to 100 R_M	Cast	<½ ½-1 1-3 >3	P P B B	8-14 6-8 3 3	4000 3500 3000 2500

P = precision   C = claw   B = buttress

Table 18.80 Guide to power band sawing

Material	Hardness	Condition	Speed, ft./min. Threads/in.				Tool mat'l
			7 or less	8-15	16-24	over 24	
Thermoplastics	All	Cast, molded, or extruded	50	50	50	50	M 10, M7, M1
Thermoplastics	All	Filled and molded	25	25	25	25	M 10, M7, M1
Thermosets	All	Cast, molded, or extruded	50	50	50	50	M 10, M7, M1
Thermosets	All	Filled and molded	25	25	25	25	M 10, M7, M1

Table 18.81 Guide to tapping TPs and TS plastics

Material	Hardness	Condition	Roughing								Finishing							
			Speed, ft./min.	Feed, in./rev. Reamer diameter, in.						Tool mat'l	Speed, ft./min.	Feed, in./rev. Reamer diameter, in.						Tool mat'l
				⅛	¼	½	1	1.5	2			⅛	¼	½	1	1.5	2	
Acrylic, acetal, polycarbonate, polysulfone, polystyrene	60 to 120 R_M	Cast, molded, or extruded	65 130	0.002 0.002	0.002 0.002	0.004 0.004	0.006 0.006	0.008 0.008	0.010 0.010	M1, M2, M7 C-2	65 130	0.002 0.002	0.002 0.002	0.004 0.004	0.006 0.006	0.008 0.008	0.010 0.010	M1, M2, M7 C-2
ABS, polyarylether, polypropylene, polyethylene, cellulose acetate	50 to 120 R_R	Cast, molded, or extruded	100 165	0.002 0.002	0.002 0.002	0.004 0.004	0.006 0.006	0.008 0.008	0.010 0.010	M1, M2, M7 C-2	100 165	0.002 0.002	0.002 0.002	0.004 0.004	0.006 0.006	0.008 0.008	0.010 0.010	M1, M2, M7 C-2
Fluoroplastics: TFE, CTFE	74 to 95 R_R	Molded or extruded	65 130	0.002 0.002	0.002 0.002	0.004 0.004	0.006 0.006	0.008 0.008	0.010 0.010	M1, M2, M7 C-2	65 130	0.002 0.002	0.002 0.002	0.004 0.004	0.006 0.006	0.008 0.008	0.010 0.010	M1, M2, M7 C-2
Polyamides (nylons): 6, 6/6, 6/12, 11, 12	78 to 120 R_R	Molded or extruded	65 165	0.002 0.002	0.002 0.002	0.004 0.004	0.006 0.006	0.008 0.008	0.010 0.010	M1, M2, M7 C-2	65 165	0.002 0.002	0.002 0.002	0.004 0.004	0.006 0.006	0.008 0.008	0.010 0.010	M1, M2, M7 C-2
Polyamides (nylons): 35% glass-reinforced, types 6, 6/6	78 to 120 R_R	Filled and molded	50 100	0.002 0.002	0.002 0.002	0.003 0.003	0.003 0.003	0.004 0.004	0.005 0.005	T15, M42 C-2	50 100	0.002 0.002	0.002 0.002	0.003 0.003	0.003 0.003	0.004 0.004	0.005 0.005	T15, M42ᶜ C-2
Polyamides (nylons): 35% glass-reinforced, types 6/10, 6/12	40 to 50 R_E	Filled and molded	45 85	0.002 0.002	0.002 0.002	0.003 0.003	0.003 0.003	0.004 0.004	0.005 0.005	T15, M42 C-2	45 85	0.002 0.002	0.002 0.002	0.003 0.003	0.003 0.003	0.004 0.004	0.005 0.005	T15, M42ᶜ C-2
Epoxy, melamine, phenolic	100 to 128 R_M	Cast or molded	65 130	0.002 0.002	0.002 0.002	0.004 0.004	0.006 0.006	0.008 0.008	0.010 0.010	M1, M2, M7 C-2	65 130	0.002 0.002	0.002 0.002	0.004 0.004	0.006 0.006	0.008 0.008	0.010 0.010	M1, M2, M7 C-2

Table 18.82 Guide to reaming TPs and TS plastics

Material	Type	Hardness							Tool
Furan, polybutadiene	Cast	40 to 100 R_n	65 / 130	0.002 / 0.002	0.002 / 0.002	0.004 / 0.004	0.006 / 0.006	0.008 / 0.008	0.010 / 0.010	M1, M2, M7 C-2
Silicone	Cast or molded	15 to 65 Shore A	45 / 85	0.002 / 0.002	0.002 / 0.002	0.004 / 0.004	0.006 / 0.006	0.008 / 0.008	0.010 / 0.010	T15, M42 C-2
Silicone, glass-filled	Filled and molded	80 to 90 R_u	45 / 85	0.002 / 0.002	0.002 / 0.002	0.003 / 0.003	0.006 / 0.006	0.004 / 0.004	0.005 / 0.005	T15, M42 C-2
Polyimide	Molded or extruded	40 to 50 R_E	65 / 130	0.002 / 0.002	0.002 / 0.002	0.004 / 0.004	0.006 / 0.006	0.008 / 0.008	0.010 / 0.010	M1, M2, M7 C-2
Polyimide, glass-filled	Filled and molded	109 to 115 R_M	45 / 85	0.002 / 0.002	0.002 / 0.002	0.003 / 0.003	0.006 / 0.006	0.004 / 0.004	0.005 / 0.005	T15, M42 C-2
Polyurethane	Cast	65 to 95 Shore A	65 / 130	0.002 / 0.002	0.002 / 0.002	0.003 / 0.003	0.006 / 0.006	0.004 / 0.004	0.005 / 0.005	M1, M2, M7 C-2
Polyurethane	Cast	55 to 75 Shore D	100 / 165	0.002 / 0.002	0.002 / 0.002	0.004 / 0.004	0.006 / 0.006	0.008 / 0.008	0.010 / 0.010	M1, M2, M7 C-2
Allyl (DAP)	Cast	95 to 100 R_M	65 / 130	0.002 / 0.002	0.002 / 0.002	0.004 / 0.004	0.006 / 0.006	0.008 / 0.008	0.010 / 0.010	M1, M2, M7 C-2

Table 18.82 Guide to reaming TPs and TS plastics *(continued)*

Material Thickness in inches (mm)	Maximum Punching Tolerances on Sheet Stock					
	Distance Between Holes				Size of Slot or Diameter of Holes	Overall Dimensions
	Under 2" (51 mm)	2" to 3" (51–76)	3" to 4" (76–102)	4" to 5" (102–127)		
Under .062" (1.6)	±.003 (±0.08)	±.004 (±0.10)	±.005 (±0.13)	±.006 (±0.15)	±.0015 (±0.038)	±.008 (±0.20)
.062" to .093" (1.6–2.4)	±.005 (±0.13)	±.006 (±0.15)	±.007 (±0.18)	±.008 (±0.20)	±.003 (±0.076)	±.010 (±0.25)
.093" to .125" (2.4–3.17)	±.006 (±0.15)	±.007 (±0.18)	±.008 (±0.20)	±.009 (±0.23)	±.005 (±0.137)	±.015 (±0.38)

Table 18.83 Guide to standard tolerances for punched holes and slots in sheet stock

Grade	Base Material	Resin	Specific Gravity	Thickness, in. (mm)	Hardness	Water Absorption	Continuous No Load Temp. F (°C)	Tensile Strength P.S.I. (MPa) LW	CW
X	Paper	Phenolic	1.36	.010–2 (0.25–51)	110	6	225 (107)	20,000 (138)	16,000 (110)
XXP	Paper	Phenolic	1.32	.015–.250 (0.38–6.35)	100	1.8	250 (121)	11,000 (75.8)	8,500 (58.6)
XXX	Paper	Phenolic	1.32	.015–2.0 (0.38–51)	110	1.4	250 (121)	15,000 (103)	12,000 (82.7)
CE	Cotton	Phenolic	1.33	.031–2.0 (0.79–51)	105	2.2	250 (121)	9,000 (62.0)	7,000 (48.2)
LE	Cotton	Phenolic	1.33	.015–2.0 (0.38–51)	105	1.95	250 (121)	12,000 (82.7)	8,500 (58.6)
AA	Asb. Fab.	Phenolic	1.70	.062–2.0 (1.3–51)	103	3.00	275 (135)	12,000 (82.7)	10,000 (69.0)
G-3	Cont. Gl.	Phenolic	1.65	.010–2.0 (0.25–51)	100	2.7	290 (143)	23,000 (158)	20,000 (138)
G-5	Cont. Gl.	Melamine	1.90	.010–3.5 (0.25–89)	120	2.7	300 (149)	37,000 (255)	30,000 (207)
G-7	Cont. Gl.	Silicone	1.69	.010–2.0 (0.25–51)	100	.55	400 (204)	23,000 (158)	18,000 (124)
G-9	Cont. Gl.	Melamine	1.90	.010–2.0 (0.25–51)	120	.80	325 (163)	37,000 (255)	30,000 (207)
G-10	Cont. Gl.	Epoxy	1.75	.010–1.0 (0.25–25)	110	.25	250 (121)	40,000 (276)	35,000 (241)
N-1	Nylon	Phenolic	1.15	.010–1.0 (0.25–25)	105	.60	250 (121)	8,500 (58.6)	8,000 (55)
GPO-1	Gl. Mat	Polyester	1.7	.062–2.0 (1.6–51)	100	1.0	250 (121)	12,000 (82.7)	10,000 (69.0)

Table 18.84   NEMA guide to standard tolerances for punched holes and slots in high-pressure composite laminated grades of sheet stock, rods, and tubes

Cutter	Line speed (m/min)	Cuts (min^{-1}) maximum	Accuracy (±mm)	Advantages	Disadvantages
Saws	150	30	0.015	Easy set-up and large capacity	Requires cleanup via air systems, etc. and uses clamping/travel table units
Guillotine	90	50	0.015	Large capacity and angle cuts	Slow blade speed; rigids need profiled bushing and/or blade; high air consumption and few cuts per minute
Flywheel	4500	12 000	0.004	High cut rates and high accuracy	Not good under 300 cuts per minute; must adjust blade rpm for speed changes; profile bushings may be needed for rigids; limited to small-angle cuts
Die-set stationary traveling	–	90	0.00	Inline finishing and high accuracy	Price, slow line speed, long set-up time, and long runs only

Table 18.85   Guide to cutting equipment capabilities

Material	Helix angle, deg.	Point angle, deg.	Clearance, deg.
Polyethylene	10–20	70–90	9–15
Rigid polyvinyl choride	27	120	9–15
Acrylic	27	120	12–20
Polystyrene	40–50	60–90	12–15
Polyamide resin	17	70–90	9–15
Polycarbonate	27	80	9–15
Acetal resin	10–20	60–90	10–15

Table 18.86   Guide to drill geometry

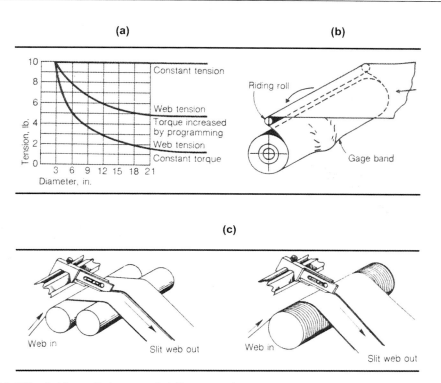

Figure 18.100 Guide to slitting extruded film or coating.

2. *Cutting, sewing, and sealing of film and sheet*. In this category of fabricating fall all the operations involved in turning TP film and sheeting into finished articles such as inflatable toys, garment bags, shower curtains, aprons, raincoats, and luggage. In making these products, the film or sheet is first cut to the desired pattern by hand, in die-cutting presses, or by other methods. The pieces are then put together using such assembly techniques as sewing, thermal methods, high-frequency sealing, or ultrasonic sealing. There is also cutting TS prepregs (B-stage cure/chapter 15) in a relatively soft form or hardened reinforced plastics (RPs). Cutting soft, precure prepregs can be done manually with scissors; it can also be automated using the machine cutting tools reviewed for the cured RPs (10, 12, 590).

3. *Forming*. In working with TP sheets, several approaches are possible. In one, the sheets can simply be bent and joined together by such techniques as hot-gas welding to form structural products such as storage tanks, hoods, venting systems, ductwork, and allied products. TP sheets can also be scored and fed into conventional paperboard

(d)

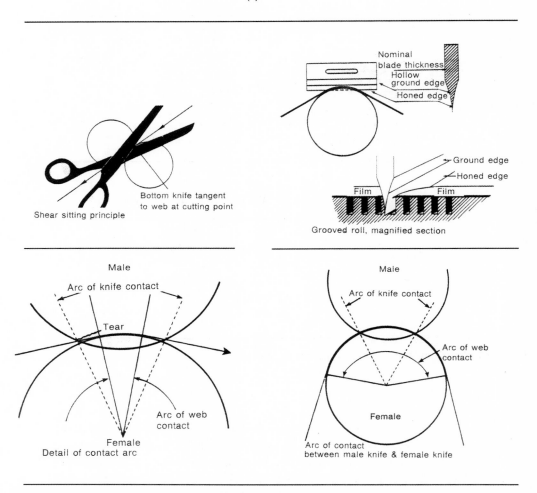

Figure 18.100 Guide to slitting extruded film or coating *(continued)*.

creasing and folding equipment to make a standard plastic box or carton. The plastic blank can also be beaded for added strength before being creased and folded into a box. The major method for forming films or sheets, however, is known as thermoforming (chapter 7).

Machining and Tooling

Consider the machining operations of turning, planing, milling, drilling, grinding, sawing, threading, tapping, reaming, polishing, and buffing. Table 18.87 and Figure 18.101 show schematic drawings of some of these tools. These examples are all ortogonal cutting processes; that is, the chip is removed at 90° to the cutting process. Figure 18.102 shows the basic mechanics of these cutting processes.

A tool of a certain rake angle (positive as shown) and relief angle moves along the surface of the workpiece at a depth of t_1. The material ahead of the tool is sheared continuously along the shear plane that makes an angle ϕ with the surface of the workpiece. This angle is called the shear angle and together with the rake angle determines the chip thickness t_2. The ratio of t_1 to t_2 is called the cutting ratio r. The relationship between the shear angle, the rake angle, and the cutting ratio is given by the equation

$$\tan \phi = r \cos \alpha / (1 - r \sin \alpha).$$

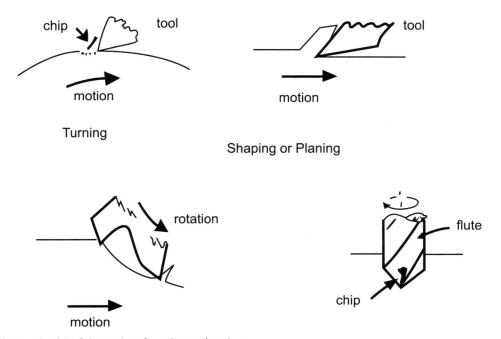

Figure 18.101 Schematics of cutting-tool actions.

TOOL STYLE	1	Conventional Turning
	2	Turning and Facing
	3	Facing

PLASTIC	A	A–3	B	C	D	E	F	G	R1	R2	R3	Surface Ft./Min.
THERMOPLASTIC	8°	8°	8°	0°	8°	20°	10°	20°	.0 to .030	.0 to .030	.093	500-900
THERMOSET	8°	8°	0°	0°	8°	20°	10°	20°	.0 to .030	.0 to .030	.093	1400-1800

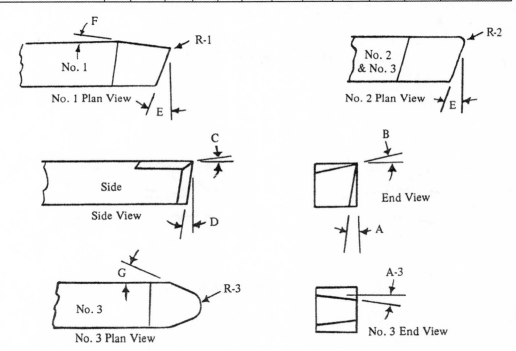

Table 18.87 Examples of cutting-tool geometries

It can be seen that the shear angle is important in that it controls the thickness of the chip. This, in turn, has a great influence on cutting performance. The shear strain that the material undergoes is given by the equation

$$\Upsilon = \cot \phi + \tan(\phi - \alpha).$$

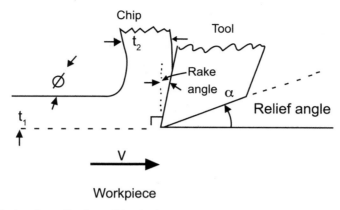

Figure 18.102 Basic schematic of a cutting tool.

Continuous chips are formed by continuous deformation of the workpiece material ahead of the tool, followed by smooth flow of the chip along the tool face. These chips are ordinarily obtained by cutting ductile materials at high speeds. Discontinuous chips consist of segments that are produced by fracture of the metal ahead of the tool. Such chips are associated with cutting brittle materials or ductile materials at very low speeds. Heterogeneous chips consist of regions of large and small strain. Such chips are characteristic of low-conductivity materials, where a large decrease in yield strength as a function of temperature is observed.

Built-up-edge chips consist of a mass of material that adheres to the tool face while the chip itself flows continuously along the face. This type of chip is associated with high friction between the chip and the tool and with poor surface finish. The forces acting on a cutting tool are shown in Figure 18.103.

The resultant force R has two components F_c and F_t. The cutting force F_c is the direction of tool travel and determines the amount of work done in cutting. The thrust force F_t does no work but together with F_c produces deflections of the tool. The resultant force also has two components on the shear plane: F_s is the force required to shear the metal along the shear plane, and F_n is the normal force on this plane. Two forces also exist on the faces of the tool, the friction force F and the normal force N. From the geometry of Figure 18.103, the following relationships may be derived, where the coefficient of friction on the face of the tool μ is given by

$$\mu = (F_t + F_c \tan \alpha)/(F_c - F_t \tan \alpha).$$

The friction force along the tool is written as

$$F = F_t \cos \alpha + F_c \sin \alpha.$$

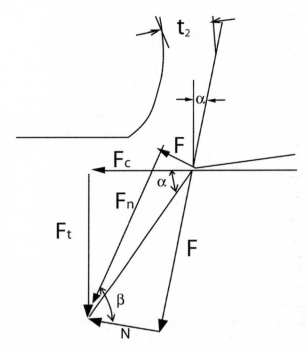

Figure 18.103 Example of forces acting on a tool.

The shear stress in the shear plane is written as

$$\tau = (F_c \sin \phi \cos \phi - F_t \sin^2 \phi)/A_o t,$$

where A_o is the cross-sectional area that is being cut. The coefficient of friction on the tool face is a complex but an important factor in cutting performance and can be reduced by such means as the use of an effective cutting fluid, higher cutting speeds, improved tool material, and conditioner chemical additives in the workpiece material. The net power consumed at the tool is calculated with the equation

$$P = F_c V/K,$$

where P is the power, F_c is the cutting force, V is the translational velocity, and K is the appropriate conversion factor.

Since F_c is a function of tool geometry, workpiece material, and process variables, it is difficult to make a reliable calculation of its value in a particular machining operation, and so in practice, empirical results in chart form are used. The data for plastics machining is still very sparse. The power consumed is then the product of unit power and the rate of metal removal.

Most of the power consumed in cutting is transformed into heat. The chip usually carries most of the heat away, but this is not often the case in plastics, owing to the low diffusion of the material. The remainder of the heat is usually divided between the tool and the workpiece; for plastics, however, the largest proportion is transmitted to the tool. A tool of high conductivity is therefore an advantage.

An increase in cutting speed or feed will, as far as metals are concerned, tend to increase the proportion of heat transferred to the chip. In plastics, this may cause degradation of the chip. This is unimportant as long as the subsequent edge melting of the chip does not foul the tool. The importance of a high-conductivity coolant is quite obvious. The cutting fluid removes the heat and thus avoids temperature buildup on the cutting edge.

A factor of great significance is tool wear. Many factors determine the type and rate at which wear occurs on the tool. The major critical variables that affect wear are tool temperature and hardness; the type of tool material; the grade and condition of the workpiece; abrasiveness in the microconstituents of the workpiece material; tool geometry; feed; cutting fluid; and surface finish of the tool. The type of wear pattern that develops depends on the relative role of these variables. Tool wear can be classified as

1. uniform abrasive wear such as that resulting in flank wear (Fig. 18.104);
2. crater wear on the tool face;
3. localized wear, such as a rounding of the cutting edge;
4. chipping of the cutting edge; and
5. concentrated wear resulting in a deep groove at the edge of the tool.

When machining plastics, a transfer film is transmitted to the tool face from the plastic material. This transfer film helps reduce the friction between the tool and the workpiece. In order to obtain the minimum friction, a surface roughness for the tool of 0.25 microns should be used. In

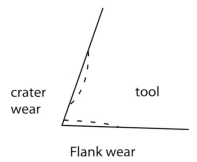

Figure 18.104 Example of wear pattern.

general, the wear on the flank or relief side of the tool is the most dependable guide for tool life. The cutting speed has the greatest influence on tool life via the equation

$$VT^n = C,$$

where V is the cutting speed, T is the cutting time (between resharpening), and n and C are constants (function of workpiece and machine variables).

Numerous investigations have been carried out modifying this equation to obtain quantitative relationships between cutting speed, feed, depth of cut, workpiece material, and tool material. The form of the results is given by the equation

$$V_1 \, d^n f^y = C_1,$$

where V_1 is the equivalent cutting speed, d is the depth of cut, f^n is the feed per revolution, and n and y are empirical exponents.

Two observations can be made from this equation. As the feed or depth of cut is increased, the cutting speed must be decreased to keep tool life constant. In doing this, the amount of material removed for the same tool life is increased considerably. Thus for a given tool life, a large amount of metal can be removed as a result of a large depth of cut and feed with a low cutting speed.

The main factors influencing surface finish in machining are

1. the outline of the cutting tool in contact with the workpiece;
2. fragments of built-up edge left on the workpiece during cutting;
3. vibration; and
4. surface degradation.

Improvements in surface finish may be obtained to various degrees by increasing the cutting speed and by decreasing the feed and depth of the cut. Changes in cutting fluid, tool geometry, and material are also important; the microstructure and chemical composition of the material also have a great influence on surface finish.

In cutting plastics, the chips may become very long and wind around the workpiece. In such cases chip breakers are used to curl the chips away from the workpiece or cause them to break into shorter sections.

Machinability is a term commonly used in machining that comprises most of the items discussed previously. This is best defined in terms of

1. tool life;
2. power requirements; and
3. surface finish.

Thus good machinability would comprise long tool life, low power requirements and good surface finish. Vibration in machine tools is often the cause of premature tool failure or short tool life, poor surface finish, damage to the workpiece, and machine damage. The vibration may be forced or self-excited (also called chatter). If the vibration is forced, it may be possible to remove or isolate the forcing element from the machine. Where the forcing frequency is near a natural frequency, either the forcing frequency or the natural frequency may be raised or lowered. Damping will also greatly reduce the amplitude.

Experience indicates that good machining practice requires a rigid setup. If a rigid setup is not available, the feed and/or the depth of cut must be reduced accordingly. Excessive tool overhang should be avoided; in milling, cutters should be mounted as close to the spindle as possible. The length of end mills and drills should be kept to a minimum. Tools with a large nose radius or with a long, straight cutting edge increase the possibility of chatter.

As a result of mechanical working and thermal effects, residual stresses exist near the surface of the workpiece. These stresses may cause warping of the material and reduce fatigue resistance. Sharp tools, medium feeds, and medium depths of cut are recommended to minimize residual stresses. The nomenclature for single-point tools, such as those used for lathes, planers, and shapers, is shown in Figure 18.105.

Each tool consist of a shank and a point. The point of a single-point tool may be formed by grinding an end of the shank. It may be forged at the end of the shank and subsequently ground; a tip or insert may be clamped or brazed to the end of the shank.

Positive rake angles improve the cutting operation with regard to forces and deflection; however, a high-positive rake angle may result in early failure of the cutting edge. Positive rake angles are generally used on lower-strength materials such as plastics. Back rake is important in plastics machining since it controls the direction of chip flow. The purpose of relief angles is to avoid interference and rubbing between the workpiece and the tool's flank surfaces. In general, they are substantial for plastics machining, but if they are too large, the tool may be weakened. The side cutting edge angle influences the length of chip contact and the time of feed. Large angles tend to cause tool chatter.

Small end cutting edge angles may create excessive force normal to the workpiece, while large angles may weaken the tool point. The purpose of the nose radius is to give a smooth surface finish and to obtain longer tool life by increasing the strength of the cutting edge. The nose radius should finish tangentially to the cutting edge's angles.

Twist drills are the most common tools used for drilling and are made in many sizes and lengths. The common nomenclature is explained in Figures 18.106. Table 18.88 provides a guide for drilling ½ to ⅜ in holes in TPs.

These drills are decreased in diameter from point to shank (back taper) to prevent binding. If the web (smallest diameter) is increased gradually in thickness from point to shank to increase strength, it is usual to reduce the helix angle as it approaches the shank. The shape of the groove is important. The shape that gives a straight cutting edge and allows a full curl to the chip is best. The helix angles vary from 0° to 45°, but the point angle is almost always 118°.

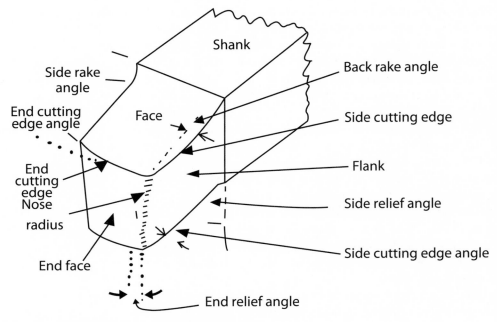

Figure 18.105 Nomenclature for single-point tools.

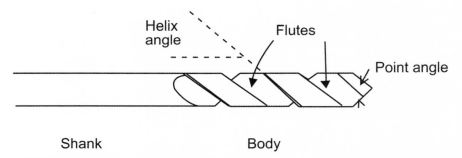

Figure 18.106 Nomenclature of twist drills.

A reamer is a tool with multiple cutting edges used to enlarge or finish holes, to give accurate dimensions as well as good finish. It is different from a drill in that it cuts only along its periphery.

Milling machines are cutters with multiple teeth, in contrast to the single-point tools of the lathe and planer. The work is generally fed past the cutter perpendicular to the cutter axis. Milling usually is face or peripheral cutting. The cutting edge has the opportunity to be cooled intermittently, as the cuts are not continuous. Milling cutters are made in a wide variety of shapes and sizes. The nomenclature is given in Figure 18.107.

Material	Drill Speed (RPM)	Feeding Speed (Low, Med., High)	Comments
Polyethylene	1000-2000	H	Easy to machine
Polyvinyl Chloride	1000-2000	M	Tend to become gummy
Acrylic	500-1500	M-H	Easy to drill with lubricant
Polystyrene	500-1500	H	Must have coolant for good hole
ABS	500-1000	M-H	
Polytetrafluoro-ethylene	1000	L-M	Easily drilled
Nylon 6/6	1000	H	Easy to drill
Polycarbonate	500-1500	M-H	Easy to drill, some gumming
Acetal	1000-2000	H	Easy to drill
Polypropylene	1000-2000	H	Easy to drill
Polyester	1000-1500	H	Easy to drill

Table 18.88 Guide for drilling ½ to ⅜ in holes in TPs

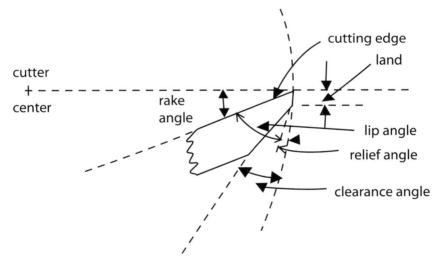

Figure 18.107 Nomenclature of milling cutters.

MACHINING NONMELT TP

PTFE cannot be processed by melt-processing techniques, such as injection molding, due to its extremely high viscosity (chapter 2). Intricate shapes required by various applications can be achieved by secondary machining of molded stock shapes or rough-molded parts. It is sometimes necessary to bond PTFE to itself or other materials. This TP is known for its nonstick properties and must be rendered adhesive using different techniques. They include pressure-sensitive adhesives or modifying the surface with sodium etching followed with an adhesive (chapter 20).

The usual high-speed machining operations (as reviewed) can be used to machine PTFE as long as the cutting tools are kept very sharp. The closest match for the free turning of fluoropolymers among metals is brass. Tool wear is similar to stainless machining. The low thermal conductivity of fluoropolymers such as PTFE causes heating of the tool and the charge material during turning. This can cause deformation of the PTFE and excessive wear of the tools. PTFE parts can be machined to a depth of less than 1.5 mm without the use of a coolant. It is necessary to use a coolant to achieve critical tolerances or to machine by automatic lathes. Dimensions of PTFE parts should be measured at a specified temperature due to the large dimensional change (1.3%) that takes place between 0°C and 100°C (146).

Standard machining operations—turning, tapping, facing, boring, drilling, threading, reaming, grinding—can be performed on PTFE and other fluoropolymers. Special machinery is not required for any of these operations. Speed selection, tool shape, and use are important considerations in successful machining. Other than poor thermal conductivity, PTFE has a coefficient of linear thermal expansion ten times higher than that of metals. This means that any type of heat buildup will cause significant expansion of the part at that point, resulting in overcuts or undercuts, thus deviating from the desired part design. Coolants should be used if surface speed of the tool exceeds 150 m/min. At higher speed, low feeds are helpful to reduce heat generation. Surface speeds between 60 and 150 m/min are satisfactory for fine-finish turning. At these speeds, feeds should be run between 0.05 to 0.25 mm/revolution. At speeds higher than 150 m/min, feeds must be dropped to a lower value.

Choice of tools is important to the control of heat buildup. Standard tools can be used but best results are obtained with specially shaped tools. A single-point tool can be designed according to the following information. The top rake should have a positive angle of 0° to 15° with a 0° to 15° side rake and side angle. The front or end rake should have an angle of 0.5° to 10°. Boring tools require angles on the high end of these ranges.

A dull tool affects tolerances achieved during machining. The tool that is not sharp pulls the stock out of alignment, thereby causing overcutting and excessive plastic removal. An improperly edged tool tends to compress, which results in shallow cuts. A very sharp tool is desirable, particularly for turning filled parts. Carbide- and Stellite-tipped tools reduce tool-sharpening frequency.

PTFE parts can be machined to have tolerances in the range of ±12 μm to ±25 μm by machining. Finishing to low tolerances is often not required for PTFE parts because they can be press fitted at a lower cost. Resilience of this plastic allows its conformation to the working dimensions. It is usually essential to relieve stress in the stock. The annealing procedure entails heating the stock shape 10°C above its service temperature (always below 327°C, which is the melting point of PTFE). It should be held at the temperature at the rate of 25 min/cm of thickness. Stresses are relieved during this operation. At the end of the hold time, the part should be cooled slowly down to room temperature. After rough cutting the part to about 300 to 500 μm, reannealing before the final cut is made will help remove stresses induced by the tool.

Lapping and grinding compounds could be used for finishing the surface of PTFE. These powders can become embedded in the surface of the part and may not be easily removed. This is true of any contaminants from machinery that is not dedicated to PTFE finishing.

Sawing and shearing

PTFE parts of any size can be sawed. Coarse saw blades are preferable to fine-toothed blades, which can become blocked with plastic. Longer saw blades perform better than a short blade, such as a hack saw. A band saw operated at moderate speeds is ideal because the long blade can remove the heat.

Shearing of rods and sheets of PTFE can be done as long as the work and the blade are firmly supported to prevent angular cuts. The limit for shearing sheets is a thickness of 10 mm and for cutting rods is a diameter of 20 mm.

Drilling, tapping, and threading

Drilling can be performed on PTFE parts using normal high-speed drills. A speed of 1000 rpm for up to 6 mm diameter and 600 rpm for up to 13 mm diameter holes are recommended. To improve the accuracy of drilling, heat should be removed. This can be accomplished by reducing friction by taking a sharply angled back-off to the cutting edge and to the polished flutes. Another consideration is the relaxation of the plastic. To increase the accuracy of the holes, a coarse hole should be drilled following a finishing cut, after allowing the part to relax for 24 hours at room temperature.

Skiving

A popular method for producing PTFE tapes is skiving a cylindrical molding of the plastic. Skiving is similar to the peeling of an apple, where a sharp blade is used at a low angle to the surface. A comparable industrial operation is the production of wood veneer.

This operation removes layers of PTFE by applying a sharp blade to the surface of the cylinder. A grooved mandrel is pressed into the center hole of the billet and the assembly mounted on a lathe. A cutting tool mounted on a rigid cross-slide is advanced toward the work at a constant speed to peel off the continuous tape with a constant thickness. The cutting tool must have an extremely sharp blade that has been finished on a fine stone and honed or stropped to prevent formation of "tram lines" on the skived film.

A typical skiving tool is depicted in Figure 18.108. Relatively thin tapes (50 to 250 μm) should be set on the center line of the stock, but for tapes with higher thickness, the blade should be set above the center line of the molding. Sheets with up to 400 μm thickness and 30 cm width can be skived on a lathe that is 20 cm wide. The blade should be fabricated from high-quality tool steel or tungsten and regularly sharpened. Skiving speed should be low and the chuck speed should be in the range of 20 to 30 rpm.

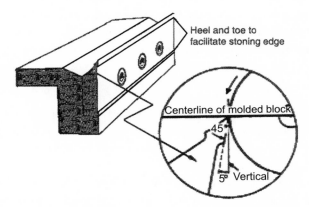

Figure 18.108 Cutting tool for machining (skiving) tape from a molded plastic block.

LASER MACHINING

Laser cutting is a fast growing process. The CO_2 laser emits radiation at a wavelength of 10.6 microns. This is in the infrared region and is strongly absorbed by plastics. If plastics were transparent to the radiation, it could not work on them. The atmosphere is quite transparent to this wavelength and so does not appreciably attenuate the beam's power. The beam can be operated usefully in the continuous-wave (CW) mode or can be pulsed to deliver powerful punches of energy. Currently, the CO_2 laser has the highest efficiency of all, converting about 10% of input power to output. This may seem low but it equals that of the electric bulb and has ten times the efficiency of other lasers.

The laser can act as a materials eliminator. Concentrating its energy on a small spot, it literally vaporizes the material in its path. If the workpiece is held stationary, the laser drills a hole. If the piece is moved, it slits the material. The induced heat is so intense and the action of the laser is so fast that only little heating of the adjacent areas of the piece takes place.

In thin materials, a single pulse may be sufficient to form the hole. Hole drilling is often done with the laser in the pulsed mode to obtain the deep, penetrating effect desired. Controlling the pulse rate and web speed to obtain the desired spacing between perforations can produce a line of perforations. The pulse lasts much less than a millisecond, thus stop-and-go motion of the web is not required. Holes can be drilled partway through an object. The beam intensity and dwell time can be adjusted to barely etch the surface or drill a hole of controlled depth.

At reduced power output and/or with defocusing of the lens, the laser can be tamed down so that it merely melts material and does not eliminate it, thus offering a scaling process. Using conventional lens systems to focus the beam, holes ranging from about 2 to 50 mils in diameter can be produced. Larger holes can be made by moving the workpiece or the laser tube in a circular fashion so as to slit the piece, much as a band saw would be used to cut a hole.

All of these effects are accomplished with no physical contact between the laser and the workpiece. The laser beam has only to focus on the area in which it is to work. Thus lasers may easily work areas accessible only with difficulty by conventional tooling and no drill chips are left behind to contaminate or scratch the material. The material removed in laser-machining operations frequently is in the form of fine dust that is removed from the area by a suction system.

Characteristic

As with any machining and fabricating technique, the laser has its own particular characteristics. For example, in conventional hole drilling, a cylindrical hole naturally results, while a conical hole is somewhat special, requiring a reaming operation with depth control of the reamer. In laser drilling, it is practically impossible to make anything other than a tapered hole and tapered sides on a cut. This action occurs because the parallel bundle of rays emitted by the laser tube must be focused to a small spot by a lens to concentrate its energy. This packet of rays between the lens and the workpiece is cone-shaped—a laser drill is literally a combination drill-reamer.

This drill-reamer may have a long taper, in which case the taper of the hole will be slight, but this long taper or long focus imposes a limit on how small a hole can be drilled. For the smallest holes, shorter focal-length lenses are used; these have a pronounced taper and so will the holes they drill. This taper is scarcely noticeable in very thin material but is more readily seen in thick material.

Other Machining Methods

Machining with a high-velocity fluid jet or hydrodynamic machining (HDM) is also applicable to many plastics. Applications range from slicing 0.75 in acoustic tile at 250 ft/min using 45000 psi to propel the jet at up to Mach 2 speeds, to shaping furniture forms of 0.5 in laminated paperboard. Shoe soles, gypsum board, urethane foam, rubber, and reinforced plastics are cut using the HDM method.

Ultrasonic machining (USM) is of particular importance when very hard materials are to be cut. HDM energy assists drilling because it can extend the drill life when producing holes in composite structures. If the plastic is conductive, electrical discharge machining (EDM) or electrochemical machining (ECM) may be useful.

Morphing, a new development in the art of automatic machining, involves the transition of one image into another. This morphing approach has been developed by computer-aided design (CAD) software maker Delcam, in Birmingham, England. It can reduce hours off lead time in the modeling or prototype process. Hybrid modeling combines solid modeling and free-form surface modeling to facilitate the modeling of complex shapes. It has the ability to change the shape of a product. With the change, all surfaces are changed (592).

Machining Safety

When the machining process generates airborne, respirable particles, there is cause for concern, regardless of the material being machined. OSHA publishes guidelines for the amount of exposure to respirable particles workers should not exceed. The list includes many of the alloying elements such as stainless steel, H13, P20 (including chromium, vanadium, nickel, copper, molybdenum, and beryllium; chapter 17). To be hazardous, these particles must be smaller than 10 μm and thus not visible to the naked eye. The large, easily seen particles or chips generated in most machining operations do not present inhalation hazards (34, 35, 97, 171, 360, 428–430, 524).

Glossary

A-, B-, and C-stages. These letters identify the various stages of cure when processing thermoset (TS) plastic that has been treated with a catalyst; basically A-stage is uncured, B-stage is partially cured, and C-stage is fully cured. The typical B-stage involves TS molding compounds and prepregs, which in turn are processed to produce C-stage fully cured plastic material products; they are relatively insoluble and infusible.

Ablative. A material that absorbs heat while part of it is being consumed by heat through a decomposition process which takes place near the surface exposed to the heat. An example is a carbon fiber-phenolic reinforced plastic (RP) that is exposed to a temperature of 1650°C (3000°F); it is the surface material of a rocket, space vehicle, and so on, designed for reentry into Earth's atmosphere from outer space.

Accuracy and repeatability. Most applications are concerned with repeatability that is easier to achieve than high accuracy. Repeatability deals with factors such as how closely the length of a given feed will repeat itself. Repeatability is different in that it does not include the noncumulative errors that an accuracy specification includes.

Adiabatic. A change in pressure or volume without gain or loss in heat. This describes a process or transformation in which no heat is added to or allowed to escape from the system.

Air shot. Also called air purge. The contents of a plasticator shot expelled into the air to study the characteristics of the melt; usually performed on start-up with the mold in the open position.

Algorithm. An algorithm is a procedure for solving a mathematical problem.

Annealing. Also called hardening, tempering, physical aging, and heat treatment. The annealing of plastics can be defined as a heat-treatment process directed at improving performance by removal of stresses or strains set up in the material during its fabrication. Depending on the plastic used, it is brought up to a required temperature for a definite time period, and then liquid (usually water or oils and waxes) and/or air cooled (quenched) to room temperature at a controlled rate. Basically the temperature is near but below the melting point. At the specified temperature the molecules have enough mobility to allow them to orient to a configuration removing or reducing residual stress. The objective is to permit stress relaxation without distortion of shape and obtain maximum performances and/or dimensional control.

Annealing is generally restricted to thermoplastics (TPs), either amorphous or crystalline. The result is increasing density, thereby improving the plastics' heat resistance and dimensional stability when exposed to elevated temperatures. It frequently improves the impact strength and prevents crazing and cracking of excessively stressed products. The magnitude of these changes depends on the nature of the plastic, the annealing conditions, and the parts' geometry.

The most desirable annealing temperatures for amorphous plastics, certain blends, and block copolymers is above their glass transition temperature (T_g) where the relaxation of stress and orientation is the most rapid. However, the required temperatures may cause excessive distortion and warping. The plastic is heated to the highest temperature at which dimensional changes owing to strain are released. This temperature can be determined by placing the plastic part in an air oven or water liquid bath and gradually raising the temperature by intervals of 3° to 5°C until the maximum allowable change in shape or dimension occurs. This distortion temperature is dictated by the thermomechanical processing history, geometry, thickness, and size. Usually the annealing temperature is set about 5°C lower using careful quality-control procedures.

Rigid, amorphous plastics such as polystyrene (PS) and acrylic (PMMA) are frequently annealed for stress relief. Annealing crystalline plastics, in addition to the usual stress relief, may also bring about significant changes in the nature of their crystalline state. The nature of the crystal structure, degree of crystallinity, size and number of spherulites, and orientation control it. In cases when proper temperature and pressure are maintained during processing, the induced internal stresses may be insignificant, and annealing is not required.

Plastic blends and block copolymers typically contain other low and intermediate molecular weight (MW) additives such as plasticizers, flame retardants, and ultraviolet (UV) or thermal stabilizers. During annealing, phase and microphase separation may be enhanced and bleeding of the additives may be observed. The morphologies of blends and block copolymers can be affected by processing and quenching conditions. If their melt viscosities are not matched, compositional layering perpendicular to the direction of flow may occur. As in the case of crystalline plastics, the skin may be different both in morphology and composition. Annealing may cause more significant changes in the skin than in the interior.

Aspect ratio. The ratio of length to diameter of a material such as a fiber or rod; also the ratio of the major to minor axis lengths of a material such as a particle. These ratios can be used in determining the effect of dispersed additive fibers and/or particles on the viscosity of a fluid/melt and in turn on the performance of the compound based on L/D ratios. In RPs, fiber L/D will have a direct influence on the RP performance.

Auger. Refers to the action of the rotating screw in advancing the plastic going from the unmelted to melted stages.

Axis. A reference line of infinite length drawn through the center of the rear of the screw shank and the center of the discharge end.

Barrel control transducer. Thermocouple and pressure transducers are inserted in different zones of the barrel to sense melt condition; they require accuracy in proper locations and recording instrumentation.

Barrel feed housing. A component of the plasticator barrel that contains the feed opening, water heating, and/or cooling channels and, in certain units, contains barrel grooving to improve the flow of plastics into the screw flights. If required, a thermal barrier is used that is attached to the barrel.

Barrel grooved feed. Grooves in the internal barrel surface in the feed section, particularly for certain materials, permits considerably more friction between the solid plastic particles and the barrel surface. The result is that process output increases and/or the process stability improves.

Beam, high performance/fire resistant. It has been known that steel structural beams cannot take the heat of a fire operating at and above 815°C (1500°F); they just loose all their strength, modulus of elasticity, and so on. To protect steel from this potential environmental condition, it can be given a temporary, short-time protection by being covered with products such as concrete and certain plastics. To significantly extend the life of structural beams, hardwood (i.e., thick wood) can be used; thus people can escape even though the wood slowly burns. The more useful and reliable structural beams should be covered in RPs that meet structural performance requirements with a much more extended supporting life than wood. However, because of RP's high costs to date, RPs are not used in this type of fire environment.

Bell end. A flange at the discharge end of the barrel that provides added strength to withstand internal pressure.

Birefringence. The difference in the refractive indexes of two perpendicular directions in a given material such as a TP. When the refractive indexes measured along three mutually perpendicular axes are identical, they are classified as optically isotropic. When the TP is stretched, providing molecular orientation and the refractive index parallel to the direction of stretching is altered so that it is no longer identical to that which is perpendicular to this direction, the plastic displays birefringence. Techniques of birefringence, ranging from the determination of structural defects in solid plastics to more basic investigations of molecular and morphological properties, are used in a wide range of applications.

Basically, birefringence is the contribution to the total birefringence of two-phase materials, due to deformation of the electric field associated with a propagating ray of light at anisotropically shaped phase boundaries. The effect may also occur with isotropic particles in an isotropic medium if they dispersed with a preferred orientation. The magnitude of the effect depends on the refractive index difference between the two phases and the shape of the dispersed particles. In TP systems the two phases may be crystalline and amorphous regions, plastic matrix and micro-voids, or plastic and filler.

Black-box. A phrase used to describe a device whose method of working is ill defined or not understood.

Blister ring. A raised portion of the root between flights of sufficient height and thickness to affect a shearing action of the melt as it flows between the blister ring and the inside wall of the barrel.

Boltzmann principle (Boltzmann, Ludwig). Ludwig Boltzmann's superposition principle provides a basis for the description of all linear viscoelastic phenomena. Unfortunately, no such theory is available to serve as a basis for the interpretation of nonlinear phenomena (i.e., to describe flows in which neither the strain or the strain rate is small). As a result, there is no general valid formula for calculating values for one material function on the basis of experimental data from another. However, limited theories have been developed.

Born in Vienna (1844–1906), Boltzmann's work of importance in chemistry became of interest in plastics because of his development of the kinetic theory of gases and rules governing their viscosity and diffusion. These are known as the Boltzmann law and Boltzmann principle, still regarded as one of the cornerstones of physical science.

British thermal unit (Btu). Btu is the energy needed to raise the temperature of 1 pound of water 1°F (0.6°C) at sea level. As an example, one pound of solid waste usually contains 4500 to 5000 Btu. Plastic waste contains greater Btu than other materials of waste.

Calculus. The mathematical tool used to analyze changes in physical quantities, comprising differential and integral calculations. It was developed during the seventeenth century to study four major classes of scientific and mathematical problems of that time: (1) find the maximum and minimum value of a quantity, such as the distance of a planet from the sun; (2) given a formula for the distance traveled by a body in any specified amount of time, find the velocity and acceleration of the body at any instant; (3) find the tangent to a curve at a point; (4) find the length of a curve, the area of a region, and the volume of a solid. These problems were resolved by the greatest minds of the seventeenth century, culminating in the crowning achievements of Gottfried Wilhelm (1646–1727) and Isaac Newton (1642–1727). Their information provided useful information for today's space travel.

Carbon black. A black colloidal carbon filler made by the partial combustion and/or thermal cracking of natural gas, oil, or another hydrocarbon.

Carbon fiber. Polyacrylonitrile (PAN) fibers are thermally carbonized to obtain carbon fiber that is used to reinforce plastics.

Catalyst. A catalyst is basically a relatively small amount of substance that augments the rate of a chemical reaction without itself being consumed, recovered unaltered in form and amount at the end of the reaction. It generally accelerates the chemical change. The materials used to aid the polymerization of most plastics are often not catalysts in the strict sense of the word (they are consumed), but common usage during the past century has applied this name to them.

There are relatively many different catalysts that are usually used for specific chemical reactions. Types include Ziegler-Natta (Z-N), metallocene, and others including their combinations. These different systems are available worldwide from different companies. Terms and information are used to identify the behavior of catalysts. An autocatalyst is a catalytic reaction induced by a product of the same reaction. This action occurs in some types of thermal decomposition. The catalyst benzoyl peroxide is a white, granular, crystalline solid; it is tasteless and has a faint odor of benzaldehyde, has active oxygen, and is soluble in almost all organic solvents. It can be used as a polymerization catalyst with different plastics such as TS polyester, rubber vulcanization without sulfur, embossed vinyl floor covering, and so on. A catalyst carrier is a neutral material used to support a catalyst, such as activated carbon, diatomaceous earth, or activated alumina. There are fluid catalysts that are finely divided solid particles utilized as a catalyst in a fluid bed process using certain TS plastics.

The enzyme catalysts are organic catalysts of living cells. Because microorganisms are able to synthesize thousands of complex organic molecules, they represent an enormous catalytic potential to the industrial chemist. A remarkable aspect of enzymes is their enormous accelerated catalytic power; they can enhance reaction rates by a factor of 10^8 to 10^{20}. Also, they can function in dilute aqueous solutions under moderate conditions of temperature and pH.

Catalyst, metallocene. Also called single-site catalyst, Me catalyst, and m catalyst. Metallocene catalysts achieve creativity and exceptional control in polymerization and molecular design, permitting penetration of new markets and expansion in present markets. Chemists can model and predict plastic structure in a matter of days rather than years. Emphasis has been on the polyolefins (mPOs); others include PS, polyethelene/polystyrene (PE/PS), thermoplastic olefins (TPO), and ethylene-propylene diene monomer (EPDM).

Uniformity of MW effectively eliminates molecular extremes resulting in a range of property improvements that are targeted to include improved mechanical, physical, and chemical properties; provide processing advantages; and lower costs. Uniquely synergistic combinations of complementary abilities are available. As an example, metallocene polyethelene (mPE) becomes an economical material competing with the properties of nylon and TP polyester plastics. Also, one can produce metallocene catalyst linear low-density polyethylene (mLLDPE) film with the same strength at a lower gauge than conventional LLDPE because of its narrow MW range. These Me catalysts are more accurate in characterizing plastics than today's quality-control instruments can verify.

They produce plastics that are stronger and tougher; thus less plastic is required. They process in a different manner so one has to become familiar with the processing techniques. The target is to obtain a plastic with a specific molecular weight distribution (MWD), density, MFR, tensile strength, flexural modulus, or a combination of other factors. Whatever the parameter, Me catalysts allow fabricators to alter reactor temperatures, pressures, and other variables to achieve their goal. Performance wise, mPO grades, regardless of density or comonomer, can combine softness and toughness, whereas conventional POs must trade off one for the other.

These catalysts can make plastics that process well by knitting long branches into the carbon chains. They make plastics with uniform, narrow MWD, high comonomer content, very even comonomer distribution, and an enormously wide choice of comonomers compared to multi-site Z-N catalysts. Comonomer choices include aromatics, styrene compounds, and cyclic olefins. Copolymers made with conventional Z-N catalysts favor ethylene and propylene. They incorporate only isolated amounts of more exotic monomers. The Me catalysts have been used to make different plastics such as PE homo-, co-, or ter-polymers from 0.865 to 0.96 density; isotactic, syndiotactic, and atactic polypropylene (PP); syndiotactic PS; and cyclic olefin copolymers.

As an example, blown film extruders designed to process LLDPE can process mLLDPE generally without difficulty; torque, head pressure, and motor load limitations generally do not limit film productivity. However, it is important to understand the differences arising from the different rheologies. The mLLDPE has a narrower MWD, and it thus exhibits lower shear sensitivity. The extruder would operate at higher temperatures and motor torque levels, while decreasing bubble stability and easing tensions on winding and draw down ratio. The Me with less chain branching would result in faster melt relaxation and less draw resonance. One with lower density would have greater elasticity, decreased specific rate in a grooved feed machine, and increased specific rate in a smoothbore machine, while harder to wind.

All other things being equal, mLLDPE is more viscous at typical extrusion shear rates than conventional LLDPE. There is a difference between shear rheology with the same screw/barrel. The mLLDPE will extrude at a higher melt temperature profile. This action may limit output on cooling-limited lines, but it may be possible to keep line speeds constant. The result is making a thinner film having the same performance because of the better properties offered by mLLDPE. Barrel cooling can be used to reduce mLLDPE melt temperatures, but it may be more desirable to optimize the extruder screw for the plastic's rheology.

Catalyst, summation. The new polymerization catalysts with conventional commodity feedstocks have produced a wave of new plastics that became obvious early during the 1990s. The terms used with this new technology include metallocene, single site, constrained geometry, and syndiotactic.

Catalyst, Ziegler-Natta. Also called the Z-N catalyst. Karl Zeigler (1898–1973) of Germany and Giulio Natta (1903–1979) of Italy developed a catalyst for the industrial production of polyolefin plastics. Together they received the Nobel Prize for chemistry in 1963. They provided the key (Zeigler for PE; Natta for PP) at that time to a relatively simple, inexpensive, and controllable large production method. They also paved the way for the overwhelming triumph of the polyolefins in subsequent years. All this is now changing to some degree, with the metallocene catalysts. Unlike the Z-N, the new generations of catalysts can provide very simplified production capabilities that produce improvements in properties, processability, and cost.

Channel. With the screw in the barrel, it is the space bounded by the interfaces of the flights, the root of the screw, and the bore of the barrel. This is the space through which the stock (melt) is conveyed and pumped.

Chisolm's law. Anytime things appear to be going better, you have overlooked something.

Chromatography. A technique for separating a sample material into constituent components and then measuring or identifying the compounds by other methods. As an example, separation, especially of closely related compounds, is caused by allowing a solution or mixture to seep through an absorbent such as clay, gel, or paper. The result is that each compound becomes adsorbed in a separate, often colored layer.

Cold flow. Creep at room temperature.

Commodity and engineering plastics. About 90wt% of plastics can be classified as commodity plastics, the others being engineering plastics. Commodity plastics are usually associated with the higher-volume, lower-priced plastics with low to medium properties. These are used for the less critical parts where engineering plastics are not required. The five families of commodities (LDPE, high-density polyethylene [HDPE], PP, polyvinyl chloride [PVC], and PS) account for about two-thirds of all the plastics consumed. The engineering plastics such as nylon, polycarbonate (PC), acetal, and so on are characterized by improved performance in higher mechanical properties, better heat resistance, higher impact strength, and so forth. Thus they demand a higher price. About a half-century ago the price-per-pound difference was at 20 cents; now it is about $1.00. There are commodity plastics with certain reinforcements and/or alloys with other plastics that put them into the engineering category. Many TSs and RPs are engineering plastics.

Computer science and algebra. The symbolic system of mathematical logic called Boolean algebra represents relationships between entities: either ideas or objects. George Boole of England formulated the basic rules of the system in 1847. Boolean algebra has been used extensively in the fields of chemistry and engineering associated with plastics and eventually became a cornerstone of computer science.

Constant lead. Also called uniform pitch screw. A screw with a flight of constant helix angle.

Creep. The time-dependent increase in strain in material, occurring under stress. Creep at room temperature is sometimes called cold flow. It is the change in the dimensions of a plastic under a given stress/load and temperature over a period of time, not including the initial instantaneous elastic deformation.

Curing. This is basically done to change the properties of a plastic material by chemical polycondensation or addition reactions; this term generally refers to the process of hardening a plastic. More specifically, it refers to the changing of the physical properties of a material by chemical reactions usually by the action of heat (includes dielectric heat, etc.) and/or catalyst with or without pressure. It is the process of hardening or solidification involving crosslinking, oxidizing, and/or polymerization (addition or condensation). The term curing, even though it is applied to TS and TP materials, is a term that refers to a chemical reaction (crosslinking) or change that occurs during its processing cycle. This reaction occurs with TS plastics or TS elastomers as well as crosslinked TPs that become TSs.

Deformation, plastic. Plastics have some degree of elasticity so as long as the plastic stretches within its elastic limit, it will eventually return to its original shape. When overstressed it reaches what is known as plastic deformation where the plastic will not return to its original shape.

Design source reduction. This generally defines the design, manufacture, purchase, or use of materials or products to reduce the amount of material used before they enter the municipal solid waste stream. Because it is intended to reduce pollution and conserve resources, source reduction should not increase the net amount or toxicity of waste generated throughout the life of the product. The US Environmental Prevention Agency (EPA) has established a hierarchy of guidelines for dealing with the solid waste situation. Their suggestions logically include source reduction, recycling, waste-to-energy gains, incineration, and landfill. The target is to reduce the quantity of trash.

Design verification (DV). Refers to the series of procedures used by the product-development group to ensure that a product design output meets its design input. It focuses primarily on the end of the product-development cycle. It is routinely understood to mean a thorough prototype testing of the final product to ensure that it is acceptable for shipment to the customers. In the context of design control, however, DV starts when a product's specification or standard has been established and is an on-going process. The net result of DV is to conform with a high degree of accuracy that the final product meets performance requirements and is safe and effective. According to standards established by ISO-9000, DV should include at least two of the following measures: (1) holding and recording design reviews, (2) undertaking qualification tests and demonstrations, (3) carrying out alternative calculations, and (4) comparing a new design with a similar, proven design.

Deviation. Refers to the variation from a specified dimension or design requirement, usually defining the upper and lower limits. The mean deviation (MD) is the average deviation of a series of numbers from their mean. In averaging the deviations, no account is taken of signs, and all deviations, whether plus or minus, are treated as positive. The MD is also called the mean absolute deviation (MAD) or average deviation (AD).

Deviation, root-mean-square (RMS). A measure of the average size of any measurable item (length of bar, film thickness, pipe thickness, coiled molecule, etc.) that relates to the degree of accuracy per standard deviation measurement.

Devolatilization (DV). An important operation in the processing of plastics into products without contaminants. Since contaminants in most cases are volatile relative to their plastics, they are removed from the condensed phase by evaporation into a contiguous gas phase. Such separation processes are commonly referred to as DV. The plastic to be devolatilized may be in the form of a melt or particulate solid. Separation is affected by applying a vacuum or by using inert substances, such as purging with nitrogen gas or steam.

Basically one or more volatile components are extracted from the plastic. It can be either in a solid or molten state. Two types of actions occur: (1) volatile components diffuse to the plastic-vapor interface (called diffusional mass transport), and (2) volatile components evaporate at the interface and are carried away (called convective mass transport). If (1) is less then (2), the process is diffusion-controlled. This condition represents most of the plastic devolatilization processes because plastic diffusion constants are usually low.

The important relationship in diffusional mass is Fick's law. It states that in diffusion the positive mass flux of component A is related to a negative concentration of ingredients. This law is valid for constant densities and for relatively low concentrations of component A in component B. The term binary mixture is used to describe a two-component mixture. A binary diffusivity constant of one component is a binary mixture.

The diffusional mass transport is driven by a concentration gradient, as described by Fick's law. This is very similar to Fourier's law, which relates heat transport to a temperature gradient. It is also very similar to Newton's law, which relates momentum transport to a velocity gradient. Because of the similarities of these three laws, many problems in diffusion are described with similar equations. Also, several of the dimensionless numbers used in heat transfer problems are also used in diffusion mass transfer problems.

Differential scanning calorimetry (DSC). A method in which the energy absorbed or produced is measured by monitoring the difference in energy input (energy changes) into the material and a reference material as a function of temperature. Absorption of energy produces an endothermic reaction; production of energy results in an exothermic reaction. DSC allows for studying the processing behavior of the melting action, degree of crystallization, and degree of cure and can be applied to processes involving a change in heat capacity such as the glass transition, loss of solvents, and so on.

Dilatant. Basically a material with the ability to increase its volume when its shape is changed, a rheological flow characteristic evidenced by an increase in viscosity with increasing rate of shear. The dilatant fluid, or inverted pseudoplastic, is one whose apparent viscosity increases simultaneously with increasing rate of shear; for example, the act of stirring instantly creates an increase in resistance to stirring.

Directional terminology. The following is a list of several terms related to directional properties:

    Anisotropic construction. A material in which the properties are different in different directions along the laminate flat plane; a material that exhibits different properties in response to stresses applied along the axes in different directions.

    Balanced construction. In woven RPs, equal parts of warp and fill fibers exist. Balanced construction refers to a material in which reactions to tension and compression loads result in extension or compression deformations only and in which flexural loads produce pure bending of equal magnitude in axial and lateral directions. It is an RP in which all laminae at angles other than 0° and 90° occur in ± pairs (not necessarily adjacent) and are symmetrical around the central line.

    Biaxial load. A loading condition in which a specimen is stressed in two different directions in its plane (i.e., a loading condition of a pressure vessel under internal pressure and with unrestrained ends).

    Bidirectional construction. An RP with the fibers oriented in various directions in the plane of the laminate usually identifies a cross laminate with the direction 90° apart.

    Isotropic construction. RPs having uniform properties in all directions. The measured properties of an isotropic material are independent on the axis of testing. The material will react consistently even if stress is applied in different directions; stress-strain ratio is uniform throughout the flat plane of the material.

    Isotropic transverse construction. Refers to a material that exhibits a special case of orthotropy in which properties are identical in two orthotropic dimensions but not the third: having identical properties in both transverse but not in the longitudinal direction.

    Nonisotropic construction. A material or product that is not isotropic—it does not have uniform properties in all directions.

    Orthotropic construction. Having three mutually perpendicular planes of elastic symmetry.

    Quasi-isotropic construction. It approximates isotropy by orientation of plies in several or more directions.

    Unidirectional construction. Refers to fibers that are oriented in the same direction, such as unidirectional fabric, tape, or laminate, often called UD. Such parallel alignment is included in pultrusion and filament winding applications.

Z-axis construction. In RP, the reference axis normal (perpendicular) to the X-Y plane (the so-called flat plane) of the RP.

Disc feeder. Horizontal, flat, grooved discs installed at the bottom of a hopper feeding a plasticator to control the feed rate by varying the disc's speed of rotation and/or varying the clearance between discs. A scraper is used to remove plastic material from the discs.

Dry cycle. The number of cycles the machine can perform in 1 minute, with a mold installed, but ignoring injection, plasticizing, and dwell time. The following phases are performed by the machine during a dry cycle rate measurement: (1) mold closing and clamping, (2) nozzle-to-mold approach, (3) nozzle retraction from mold, and (4) mold opening.

Elastomer. A rubberlike material (natural or synthetic) that is generally identified as a material that at room temperature stretches under low stress to at least twice its length and snaps back to approximately its original length on release of the stress (pull) within a specified time period. The term elastomer is often used interchangeably with the term plastic or rubber; however, certain industries use only one or the other.

Although rubber originally meant a natural TS elastomeric material obtained from a rubber tree (*Hevea braziliensis*), it identifies a thermoset elastomer (TSE) or thermoplastic elastomer (TPE) material. They can be differentiated by how long a deformed material requires to return to its approximately original size after the deforming force is removed and by its extent of recovery. Different properties also identify the elastomers such as strength and stiffness, abrasion resistance, solvent resistance, shock and vibration control, electrical and thermal insulation, waterproofing, tear resistance, cost to performance, and so on.

The natural rubber materials have been around for over a century. They will always be required to meet certain desired properties in specific products. TPEs principally continue to replace traditional TS natural and synthetic rubbers (elastomers). TPEs are also widely used to modify the properties of rigid TPs, usually by improving their impact strength.

Natural rubber provides the industry worldwide with certain material properties that to date are not equaled by synthetic elastomers. Examples of some of the products produced include tires (with heat build-up resistance), certain type vibrators, and so on. However, both synthetic TSE and TPE have made major inroads to product markets previously held by natural rubber and have also expanded into new markets. The three basic processing types are conventional (vulcanizable) elastomer, reactive type, and TPE.

Electrical corona discharge treatment. A method for rendering inert plastics, such as polyolefins, more receptive to inks, adhesives, or decorative coatings by subjecting their surfaces to a corona discharge. A typical method of treating films is to pass the film over a grounded metal cylinder that is located above a sharp-edged high-voltage electrode spaced so as to leave a small gap.

The corona discharge oxidizes the film by means of the formation of polar groups on reactive sites making the surface receptive to coatings and so on.

Endotherm. A process or change that takes place with absorption of heat and requires high temperature for initiation and maintenance, as with using heat to melt plastics and then removing heat; as opposed to endothermic.

Endothermic. Also called endoergic. Pertaining to a reaction that absorbs heat.

Energy. Basically, it is the capacity for doing work or producing change. This term is both general and specific. Generally it refers to the energy absorbed by any material subjected to loading. Specifically it is a measure of toughness or impact strength of a material (e.g., the energy needed to fracture a specimen in an impact test). It is the difference in kinetic energy of the striker before and after impact, expressed as total energy per inch of notch of the test specimen for plastic and electrical insulating material [in-lb (J/m)]. Higher energy absorption indicates a greater toughness. For notched specimens, energy absorption is an indication of the effect of internal multiaxial stress distribution on fracture behavior of the material. It is merely a qualitative index and cannot be used directly in design.

Energy and bottle. An interesting historical (1950s) example is the small-injection blow-molded whiskey bottles that were substituted for glass-blown bottles in commercial aircraft; these continue to be used in all worldwide flying aircraft. At that time, just in United States, over 500×10^{12} Btu, or the amount of energy equivalent to over 80×10^6 barrels of oil, was reduced per year.

Enthalpy. It refers to the quantity of heat, equal to the sum of the internal energy of a system plus the product of the pressure-volume work performed on the system such as the action during heat processing of plastics. As a thermodynamic function, it is defined by the equation $H = U + PV$, where H = enthalpy, U = internal energy, P = pressure, and V = volume of the system.

Entropy. A measure of the unavailable energy in a thermodynamic system, commonly expressed in terms of its exchanges on an arbitrary scale with the entropy of water at 0°C (32°F) being zero. The increase in entropy of a body is equal to the amount of heat absorbed divided by the absolute temperature of the body.

Euler equation. A special case of the general equation of motion. It applies to the flow systems in which the viscous effects are negligible.

Eutectic blend. A mixture of two or more substances that solidifies as a whole when cooled from the liquid state, without changing composition. It is the composition within any system of two or more crystalline phases that melts completely at the minimum temperature.

Exotherm. The temperature versus the time curve of a chemical reaction or a phase change giving off heat, particularly the polymerization of TS; the heat liberated by chemical reactions accelerated during processing. Maximum temperature occurs at peak exotherm. Some plastics such as room temperature curing TS polyesters and epoxies will exotherm severely with damaging results if processed incorrectly. As an example, if too much methyl ethyl ketone peroxide (MEK peroxide) catalyst is added to polyester plastic that contains cobalt naphthenate (promoter), the mix can get hot enough to smoke and even catch fire. Thus an exotherm can be a help or hindrance, depending on the application (e.g., during casting, potting, etc.).

Extruder, adiabatic. Also called autothermal. Describes a process or transformation in which no heat is added to or allowed to escape from the system under consideration. It is used, somewhat incorrectly, to describe a mode of a process such as an extruder in which no external heat is added to the extruder. Although heat may be removed by cooling, this keeps the output temperature of the melt passing through the extruder at a constant and control rate. The screw develops the heat input in such a process as its mechanical energy is converted to thermal energy.

Extruder, autogenous. Some extruders operate without forced cooling or heating. This is the so-called autogenous extrusion operation; it is not to be confused with an adiabatic extruder. An autogenous process is where the heat required is supplied entirely by the conversion of mechanical energy into thermal energy. However, heat losses can occur in an autogenous process. An adiabatic process is one where there is absolutely no exchange of heat with the surroundings. An autogeneous extrusion operation can never be truly adiabatic, only by approximation.

In practice, autogeneous extrusion does not occur often because it requires a delicate balance between plastic properties, machine design, and operating conditions. A change in any of these factors will generally cause a departure from autogeneous conditions. The closer one operates to autogeneous conditions, the more likely it is that cooling will be required. Given the large differences in thermal and rheological properties of plastics, to date it is difficult to design an extruder that can operate in an autogeneous fashion with several different plastics. Therefore, most extruders are designed to have a reasonable amount of energy input from external barrel heaters.

Extruder, isothermal. A process where the melt stocks remain constant for a good portion in the plasticator. This type of operation is most common in small diameter screw extruders.

Fatigue. The action that causes a failure or deterioration in mechanical properties after repeated, cyclic applications of stress. Test data provides information on the ability of a material to resist the developments of cracks, which eventually bring about failure as a result of long periods of the cyclic loading.

Feed side opening. An opening that feeds the material at an angle into the side of the screw rather than the more conventional system of feeding vertically downward on the screw.

Finagle's law. Once a job is fouled up, anything done to improve it makes it worse.

Flight land. The surface of the radial extremity of the flight constituting the periphery or outside diameter of the screw.

Foamed plastic. Practically all plastics can be made into foams. When compared to solid plastics, density reduction can go from near solid to almost a weightless plastic material. There are so-called plastic structural foams (SFs) that have up to 40% to 50% density reduction. The actual density reduction obtained will depend on the product's thickness, the product's shape, and the melt flow distance during processing (e.g., how much plastic occupies the mold cavity).

Ford car. The gasoline-powered automobile was not invented by Henry Ford. It was independently developed by Gottlieb Daimler and Karl Benz in the last decade of the nineteenth century. Several years later, Henry Ford invented the moving assembly line. That flash of brilliance was the means of producing cars cheaply and in great numbers.

Fossil fuels. Fossil fuels (coal, crude oil or petroleum, natural gas liquids, and natural gas) are the primary sources of basic petrochemicals. About 3% are used to produce plastic materials. The most important use that consumes most of the fossil fuels is in the production of energy.

Fuzzy logic control (FLC). Although FLC may sound exotic, it has been used to control many conveniences of modern life (from elevators to dishwashers) and, more recently, in the industrial process control that includes plastic processing such as temperature and pressure. FLC actually outperforms conventional controls because it completely avoids overshooting process limits and dramatically improves the speed of response to process upsets. These controllers accomplish both goals simultaneously, rather than trading one against another as done with proportional-integral-derivative (PID) control. However, FLC is not a cure all because not all FLCs are equal—no more than PIDs. FL is not needed in all applications; in fact, FLCs allow themselves to be switched off so that traditional PID control takes over.

Galois, Evariste. Evariste Galois, now recognized as one of the greatest nineteenth century mathematicians, failed the entry exam for the Ecole Polytechnique twice, and a paper he submitted to the French Academy of Sciences was rejected as "incomprehensible." Embittered, he turned to political activism and spent 6 years in prison. In 1832, at the age of 20, he was killed in a duel, reported to have arisen from a lover's quarrel (although there were those who believed that an agent provocateur of the police was involved).

Glass transition temperature. Also called glass-rubber transition; identified as T_g. Basically this important characteristic is the reversible change in phase of a plastic from a viscous or rubbery state to a brittle glassy state. T_g is the point below which plastic behaves like glass while still remaining very strong and rigid. Above this temperature, it is not as strong or rigid as glass but it is not brittle. At T_g the plastic's volume or length increases, and above it properties decrease. The amorphous TPs have a more definite T_g when compared to crystalline plastics. It is usually reported as a single value. However, it occurs over a temperature range and is kinetic in nature. An example of the T_g range has PE at −125°C and PMMA at +105°C.

Hub. The portion immediately behind the flight that prevents the escape of the plastic. A sealing device is used to prevent leakage of plastic back around the screw hub, usually attached to the rear of the feed section.

Hysteresis effect. The hysteresis effect is a retardation of the strain when a material is subjected to a force or load. Recovery of strain in a material is subjected to a stress during its unloading cycle due to energy consumption. This energy is converted from mechanical to frictional energy (heat). It can represent the difference in a measurement signal for a given process property value when approached first from a zero load and then from a full scale.

Inching. A reduction in the rate of mold-closing travel just before the mating mold surfaces touch each other.

Isotactic molding. Also called isotactic pressing or hot isotactic pressing (HIP). The compressing or pressing of powder material (e.g., plastic, etc.) under a gas or liquid so the pressure is transmitted equally in all directions. Examples include autoclave, sintering, injection-compression molding, elastomeric mold using hydrostatic pressure, and underwater sintering.

Jetting. An undesirable melt entering the cavity, rather than being in a parabolic melt front; the melt squirts through the gate into the cavity in a worm- or a snake-like pattern. Causes of this include an undersized gate or thin to thick cavity section resulting in poor control of the molded part.

Kinetic. A branch of dynamics concerned with the relations between the movement of bodies and the forces acting upon them.

Kinetic theory. A theory of matter based on the mathematical description of the relationship between pressures, volumes, and temperatures of gases (PVT phenomena). This relationship is summarized in the laws of Boyle's law, Charles's law, and Avogadro's law.

Liquid crystalline polymer (LCP). LCPs are best thought of as being a separate, unique class of TPs. Their molecules are stiff, rod-like structures organized in large parallel arrays or domains in both the melted and solid states. These large, ordered domains provide LCPs with characteristics that are unique compared to those of the basic crystalline or amorphous plastics. They are called self-reinforcing plastics because of their densely packed fibrous polymer chains.

Logarithm. It is the exponent that indicates the power to which a number is raised to produce a given number. Thus, as an example, 1000 to the base of 10 is 3 (since $10^3 = 1000$). This type of mathematics is used extensively in computer software.

Mathematical dimensional eccentricity. The ratio of the difference between maximum and minimum dimensions on a product, such as wall thickness. It is expressed as a percentage to the maximum.

Mathematical tool. Calculus is the mathematical tool used in plastic R&D programs to analyze changes in physical quantities, comprising differential, integral calculations, and so on. It was developed during the seventeenth century to study four major classes of scientific and mathematical problems of that time: (1) find the maximum and minimum value of a quantity, such as the distance of a planet from the sun; (2) given a formula for the distance traveled by a body in any specified amount of time, find the velocity and acceleration of the body at any instant; (3) find the tangent to a curve at a point; (4) find the length of a curve, the area of a region, and the volume of a solid. These problems were resolved by the greatest minds of the seventeenth century, culminating in the crowning achievements of Gottfried Wilhelm (1646–1727) and Isaac Newton (1642–1727). Their information provided useful information for today's space travel.

Mean. Arithmetical average of a set of numbers. It provides a value that lies between a range of values and is determined according to a prescribe law.

Mean absolute deviation (MAD). MAD is a statistical measure of the mean (average) difference between a product's forecast and actual usage (demand). The deviations (differences) are included without regard to whether the forecast was higher or lower than the actual usage.

Meld line. Refers to a line that is similar to a weld line, except that the flow fronts move parallel rather than meeting head on.

Melt. Plastic in a molten or plasticated condition. This also refers to the an extruder's extrudate.

Melt deformation. As a melt is subjected to a fixed stress or strain, the deformation versus time curve will show an initial rapid deformation followed by a continuous flow. When elasticity and strain are compared they provide (1) basic deformation versus the time curve, (2) stress-strain deformation versus time with the creep effect, (3) stress-strain deformation versus time with the stress-relaxation effect, (4) material exhibiting elasticity, and (5) material exhibiting plasticity. The relative importance of elasticity (deformation) and viscosity (flow) depends on the time scale of the deformation. For a short time elasticity dominates, but over a long time the flow becomes purely viscous. This behavior influences processes.

Deformation contributes significantly to process-flow defects. Melts with only small deformation have proportional stress-strain behavior. As the stress on a melt is increased, the recoverable strain tends to reach a limiting value. It is in the high-stress range, near the elastic limit, that processes operate.

MW, temperature, and pressure have little effect on elasticity; the main controlling factor is MWD. Practical elasticity phenomena often exhibit little concern for the actual values of the modulus and viscosity. Although MW and temperature influence the modulus only slightly, these parameters have a great effect on viscosity and thus can alter the balance of a process.

Melt fracture. Also called elastic turbulence. It is the instability or an elastic strain in the melt flow usually through a die starting at the entry of the die. It leads to surface irregularities on the finished part like a regular helix or irregularly spaced ripples. Plastic's rheology influences its melt fracture behavior. Higher MW plastics (with MWD) tend to have less sensitivity to its onset. This fracture can also occur in molds with complex cavities and/or improper melt flow with in the mold.

Melt index. A term used that indicates how much plastic melt can be pushed through a set orifice with various conditions controlled (basically temperature, time, pressure). It represents the "flowability" of a material. The higher numbers indicate the easier flow.

Modulus of elasticity. Most materials, including plastics and metals, have deformation proportional to their loads below the proportional limits. (A material's proportional limit is the greatest stress at which it is capable of sustaining an applied load without deviating from the proportionality of a stress-strain straight line.) Since stress is proportional to load and strain to deformation, this implies that stress is proportional to strain. Hooke's law, developed in 1676, follows that this straight line of proportionality is calculated as stress/strain = constant. The constant is called the modulus of elasticity or Young's modulus (defined by Thomas Young in 1807, although others used the concept, including the Romans and the Chinese [BCE]).

Molecular weight (MW). MW is the sum of the atomic weights of all the atoms in a molecule. It represents a measure of the chain length for the molecules that make up the polymer. Atomic weight is the relative mass of an atom of any element based on a scale in which a specific carbon atom is assigned a mass value of 12.

The MW of plastics influences their properties. As an example, with increasing MW, properties increase for abrasion resistance, brittleness, chemical resistance, elongation, hardness, melt viscosity, tensile strength, modulus, toughness, and yield strength. Decreases occur for adhesion, melt index, and solubility.

Adequate MW is a fundamental requirement to meet desired properties of plastics. With MW differences of incoming material, the fabricated product performance can be altered; the more the difference, the more dramatic change occurs in the product. Melt flow rate (MFR) tests are used to detect degradation in products where comparisons, as an example, are made of the MFR of pellets to the MFR of products. MFR has a reciprocal relationship to melt viscosity. This relationship of MW to MFR is an inverse one; as the MW drops, the MFR increases the MW. The melt viscosity is also related: as one increases the other increases.

The average molecular weight (AMW) is the sum of the atomic masses of the elements forming the molecule, indicating the relative typical chain length of the polymer molecule. Many techniques are available for its determination. The choice of method is often complicated by limitations of the technique as well as by the nature of the polymer because most techniques require a sample in solution.

Molecular weight distribution (MWD). MWD is basically the amount of component polymers that go to make up a polymer. *Component polymers*, in contrast, is a convenient term that recognizes the fact that all polymeric materials comprise a mixture of different polymers of differing MWs. The ratio of the weight AMW to the number average MW gives an indication of the MWD. AMW information is useful; however, characterization of the breadth of the distribution is usually more valuable. For example, two plastics may have exactly the same or similar AMWs but very different MWDs. There are several ways to measure MWD, such as the fractionation of a polymer with broad MWD into narrower MWD fractions.

Monocoque structure. Plastics provide an easy means of producing monocoque constructions, such as in different applications that include aircraft fuselage, automotive body, motor truck, railroad car, and houses. Its construction is one in which the outer-covering "skin" carries all or a major part of the stresses. The structure can integrate its body and chassis, such as in aircraft and automobiles.

Monomer. Plastics are predominantly organic (carbon containing) compounds primarily made up of six elements forming a monomer such as ethylene that is a gas. Another example of a monomer is vinyl chloride that is also a gas.

Morphology. The study of the physical form or structure of a material; the physical microstructure of a bulk polymer. Common units are lamella, spherulite, and domain. In turn there are TPs and TS plastics. Lamella is a thin, flat scale layer of polymers. Spherulite is a rounded aggregate of radiating lamellar crystals. Spherulites exist in most crystalline plastics and usually impinge on one another to form polyhedrons. They range in size from a few tenths of a micron in diameter to several millimeters. Domain is a microphase of one polymer in a multiphase system.

Nanocomposite. Plastics derived from compounding nanofillers (clays and other particles) in polymers.

NEAT plastic. Identifies a plastic with "nothing else added to" it. It is a true virgin polymer since it does not contain additives, fillers, and so on. These are rarely used.

Newtonian flow. A flow characteristic where a material (e.g., a liquid) flows immediately on application of force and for which the rate of flow is directly proportional to the force applied. It is a flow characteristic evidenced by viscosity that is independent of shear rate. Water and thin mineral oils are examples of Newtonian flow.

Nonlaminar flow. Ideally, it is a melt flow in a steady, streamlined pattern in and/or out of a tool (e.g., die, mold, etc.). Usually, the melt is usually distorted causing defects called melt fracture or elastic turbulence. To reduce or eliminate this problem, the entrance to the die or mold is tapered or streamlined.

Non-Newtonian flow. A flow characteristic where materials such as plastic have basically abnormal flow response when force is applied (i.e., their viscosity is dependent on the rate of shear). They do not have a straight proportional behavior with application of force and rate of flow. When proportional, the behavior has a Newtonian flow.

Orientation, balance. It is the result where stretch in the machine and transverse directions are uniform.

Orientation, biaxial. Also called biorientation or BO. It is the stretching of material in two directions (biaxially) at right angles: along machine direction (MD) and across or transverse direction (TD). The difference in the amount of stretch in both directions varies depending on product requirement. If they are equal, then it is a balanced orientation. Small to large size lines are used. One of the largest cast-oriented polyethylene terephthalate (PET) film lines in the world (DuPont's plant in Dumphries, Scotland built by Kampf GmbH and Company, Germany) produces 9-meter-wide film after MD and TD stretching. The film is wound in one piece at up to 480 m/min. A take-up roll weigh is 13 tons using a high stiffness RP carbon fiber/epoxy core. There is also a very large polypropylene plastic coextruded film line (Applied Extrusion Technologies Inc., New Castle, Delaware). It uses a massive tenter oven and turret winder (built by Bruckner Maschinenbau, Germany) that produces 10-meter-wide film at up to 400 m/min or 50 million lb/yr.

Orientation, cold stretching. Plastics may be oriented by the so-called cold stretching; that is, below its glass transition temperature (T_g). There has to be sufficient internal friction to convert mechanical into thermal energy, thus producing local heating above T_g. This occurs characteristically in the necking of fibers during cold drawing.

Orientation and glassy state. An important transition occurs in the structure of both crystalline and noncrystalline plastics. This is the point at which they transition out of the so-called glassy state. Rigidity and brittleness characterize the glassy state. This is because the molecules are too close together to allow extensive slipping motion between each other. When the glass transition (T_g) is above the range of the normal temperatures to which the part is expected to be subjected, it is possible to blend in materials that can produce the T_g of the desired mix. This action yields more flexible and tougher plastics.

Orientation and heat-shrinkability. There are oriented heat-shrinkable plastic products found in flat, tubular film and tubular sheet. The usual orientation is terminated (frozen) downstream of a stretching operation when a cold enough temperature is achieved. Reversing this operation occurs when the product is subjected to a sufficient high temperature. This reheating results in the product shrinking. Uses for these products include part assemblies, tubular or flat communication cable wraps, furniture webbing, medical devices, wire and pipe fitting connections or joints, and so on.

Orientation and mobility. Orientation requires considerable mobility of large segments of the plastic molecules. It cannot occur below the glass transition temperature (T_g). The plastic temperature is taken just above T_g.

Orientation tenter mark. A visible deformation on the side edge of a material due to the pressure from the clips and clamps; this trim is cut.

Orientation, thermal characteristic. These oriented plastics are considered permanent, heat-stable materials. However, the stretching decreases dimensional stability at higher temperatures. This situation is not a problem since these types of materials are not exposed to the higher temperatures in service. For the heat-shrink applications, the high heat provides the shrinkage capability.

Orientation, uniaxial. Also called axial orientation, monoaxial orientation, or UO. This is the stretching only in one direction that is usually in the machine direction.

Orientation, wet stretching. For plastics whose glass transition temperature (T_g) is above their decomposition temperature, orientation can be accomplished by swelling them temporarily with plasticizing liquids to lower their T_g of the total mass, particularly in solution processing. As an example, cellulose viscous films can be drawn during coagulation. Final removal of the solvent makes the orientation permanent.

pH. An expression of the degree of acidity or alkalinity of a substance. With neutrality being at pH 7, acid solutions are less than 7 and alkaline solutions are more than 7.

Pitch. The high MW residue from the destructive distillation of petroleum and coal products. They can be used as base materials for the manufacture of high-modulus carbon fibers.

Plastic volume swept. The volume of material that is displaced as the screw (or plunger) moves forward. It is the effective area of the screw multiplied by the distance of travel.

Plasticator. A very important component in machines for extrusion, injection molding, and blow molding by providing a melting process via barrel and screw. If factors such as the proper screw design or barrel heat profile are not considered, products may not meet or maximize their performance or meet their low cost requirements.

Plasticity. The inverse condition of elasticity. The material tends to stay in its deformed shape, which often occurs when it is stressed beyond its yield point.

Pocket. A place where a screw flight is initiated, usually starting from a cylindrical area or another flight. The feed pocket exists on most screws and is located at the intersection of the bearing and the beginning of the flight.

Poisson's ratio. It is the proportion of lateral strain to longitudinal strain under conditions of uniform longitudinal stress within the proportional or elastic limit. When the material's deformation is within the elastic range, it results in a lateral to longitudinal strain that will always be constant. In mathematical terms, Poisson's ratio is the diameter of the test specimen before and after elongation divided by the length of the specimen before and after elongation. Poisson's ratio will have more than one value if the material is not isotropic.

Polymer. When monomers are basically subjected to a catalyst, heat, and/or pressure, the double bonds open up and the individual monomer units join "arms" to form long chains called polymers. This process is called polymerization. Polymerization is basically the bonding of many monomer units to produce polymers.

There is a type of chemical reaction (addition or condensation) in which the molecules of a monomer are linked together to form large molecules whose MW is a multiple of that of the original substance resulting in high MW components.

The geometry of these chains is just as important as their chemical makeup in determining plastic processability and properties of fabricated products. Chains can be long, having thousands of repeating monomer units. Short chains have fewer repeating monomer units.

Polyolefin plastic. Also called olefin, olefinic plastics, or olefinic resins. They represent a very large class of carbon-chain TPs and elastomers. The most important are polyethylenes and polypropylenes. They all have extensive use in many different forms and applications.

Postconsumer. Identifies plastic products generated by a business or consumer that have served their intended purpose and that have been separated or diverted from solid waste for the purposes of collection, recycling, and/or disposition

Preconsumer. Any material that has not made its way to the consumer. It includes scrap, waste, and rejected parts or products.

Prepreg. Term generally used in RPs for a reinforcement containing or combined usually with a TS liquid plastic (TPs are also used) that can be stored under controlled conditions. Reinforcement (e.g., fibers and/or rovings, woven and/or nonwoven fabrics, etc.) can be in different forms and patterns. The TS is completely compounded with catalysts and so on and partially cured to the required tack state in the B-stage. The fabricator completes the cure with heat and pressure.

Pressure, atmospheric. The atmosphere (atm) is the envelope of gases (air) that surrounds Earth exerting pressure on Earth with certain plastic-fabricating processes taking advantage of this pressure. At various altitudes in feet, in approximate absolute pressure in psia (gauge in of Hg), they are: sea level at 14.7 (0.0), 1000 at 14.2 (1.0), 2000 at 13.7 (2.1), 2000 at 13.2 (3.1), 4000 at 12.7 (4.1), 5000 at 12.2 (5.0), 6000 at 11.7 (6.0), and so on. The pressure exerted at sea level is 14.696 psi (101.325 kPa), which will support a column of mercury (Hg) 760 mm high (about 30 in), having a density of 13.5951 g/cm^3 at a temperature of 0°C (32°F) and standard gravity of 980.665 cm/s^2. This atm is a standard barometric pressure, although it varies slightly with local meteorological conditions. This pressure is used in fabricating processes where only contact or very low pressure such as vacuum pressure (where atmospheric pressure is applied) is required. Those processes include certain casting, coating, and RP systems.

Processing. The art of processing plastic is an "art of detail." The more you pay attention to details, the fewer hassles you will get from the process. Note that if it has been running well, it will continue running well unless a change occurs. Correct the problem; do not compensate. It may not be an easy task, but understanding what you have equipment wise, material wise, environment wise, and/or people wise can do it.

Processing, feedback. The information returned to a control system or process to maintain the output within specific limits.

Processing, in-line. A complete production or fabricating operation can go from material storage and handling to produce the part, which includes upstream and downstream auxiliary equipment. It then passes through inspection and quality control to packaging and is delivered to destinations such as warehouse bins or transportation vehicles.

Processing, intelligent. What is needed is to cut inefficiency, such as the variables, and in turn cut the costs associated with them. One approach that can overcome these difficulties is called intelligent processing of materials. This technology utilizes new sensors, expert systems, and process models that control processing conditions as materials are produced and processed without the need for human control or monitoring. Sensors and expert systems are not new in themselves. What is novel is the manner in which they are tied together. In intelligent, new, nondestructive evaluation, sensors are used to monitor the development of a materials microstructure as it evolves during production in real time. These sensors can indicate whether the microstructure is developing properly. Poor microstructure will lead to defects in materials. In essence, the sensors are inspecting the material on-line before the product is produced.

Processing line, downstream. The plastic discharge end of the fabricating equipment such as the auxiliary equipment in an extrusion pipeline after the extruder.

Processing line, downtime. Refers to equipment that cannot be operated when it should be operating. Some reasons for downtime could involve equipment that is inoperative, a shortage of material, an electric power problem, a lack of available operators, and so on. Regardless of reason, downtime is costly.

Processing line, upstream. Refers to material movement and auxiliary equipment (e.g., dryers, mixers/blenders, storage bins, etc.) that exist prior to plastic entering the main fabricating machine such as the extruder.

Processing line, uptime. When the plant is operating to produce products.

Processing parameter. Measurable parameters such as temperature and pressure required during preparation of plastic materials, during processing of products, inspection, and so on.

Processing stabilizer. Also called a flow promoter. In TPs they act in the same manner as internal lubricants, where they plasticize the outer surfaces of the plastic particles and ease their fusion, but can be used in greater concentrations (about 5 pph). With TS plastics they are not reactive normally, which therefore reduces the rate of interactions of reactive groupings by a dilution effect. Thus easier processing may be derived mainly from the reduction in the rate at which the melt viscosity increases. At the same time, the overall crosslinking density is reduced.

Processing via fluorescence spectroscopy. Sensor techniques can measure the properties of plastics during processing. The intent is to improve product quality and productivity by using molecular or viscous properties of the melt as a basis for process control, replacing the indirect variables of temperatures, pressure, and time. This system analyzes the fluorescence generated in the plastic during processing and translates it into a numerical value for the property being monitored. The plastic must be doped with a small amount of fluorescence dye specific to the application. An optical fiber installed in the plasticator barrel, mold, or die scans the plastic. It is used to perform other tasks such as measuring the concentration and dispersion uniformity of filler; accuracy of 1% provides a means of optimizing residence time. It can also monitor the glass transition temperature.

Qualified products list (QPL). QPL is a list of commercial products that have been pretested and found to meet the requirements of a specification.

Relief. An area of the screw shank of lesser diameter than the outside diameter and located between the bearing and the spine or key way.

Residence time. The amount of time a plastic is subjected to heat during fabrication of virgin plastics, such as during extrusion, injection and compression molding, calendering, and so on. With recycled plastics, properties are affected by previous fabrication and granulating heat. This residence time can cause relatively minor to definite major undesirable effects or variations in properties of the plastic during the next processing step and/or the finished product. This action can occur even when the same plastic (from the same source) and same fabricating machine are used. Different thermal tests are available and used to meet specific requirements.

Residual stress. The stress existing in a body at rest, in equilibrium, at uniform temperature, and not subjected to external forces. Often caused by the stresses remaining in a plastic part as a result of thermal and/or mechanical treatment in fabricating parts. Usually these stresses are not a problem in the finished product. However, with excess stresses, the product could be damaged quickly or after being in service from a short to long time, depending on amount of stress and the environmental conditions around the product.

Reynolds number. A dimensionless number that is significant in the design of any system in which the effect of viscosity is important in controlling the velocities or the flow pattern of a fluid. It is equal to the density of a fluid times its velocity, times a characteristic length, divided by the fluid viscosity. This value or ratio is used to determine whether the flow of a fluid through a channel or passage, such as in a mold, is laminar (streamlined) or turbulent.

Rifled liner. The barrel liner whose bore is provided with helical grooves.

Risk. Designers and others in the plastics industry and other industries have the responsibility to ensure that all products produced will be safe and not contaminate the environment. Recognize that when you encounter a potential problem, you are guilty until proven innocent (or is it the reverse?). So keep the records you need to survive the legal actions that can develop.

Risk, acceptable. This is a concept that developed decades ago in connection with toxic substances, food additives, air and water pollution, fire and related environmental concerns, and so on. It can be defined as a level of risk at which a seriously adverse result is highly unlikely to occur but it cannot be proven whether or not there is 100% safety. In these cases, it means living with a reasonable assurance of safety and acceptable uncertainty.

Shank. It is the rear protruding portion of the plasticator screw to which the driving force is applied.

Skiving. A specialized process for producing film is skiving. It consists of shaving off a thin film or sheet layer from a large block of solid plastic such as a round billet. Continuous film is obtained by skiving in a lathe-type cutting operation that is similar to producing plywood from a tree-trunk log. This process is particularly useful with plastics that cannot be processed by the usual plastic film processes, such as extrusion, calendering, or casting. PTFE is an example, as it is a plastic that is not basically melt processable.

Static mixer. Also called a motionless mixer. They are designed to achieve a homogeneous mix by flowing one or more plastic streams through geometric patterns formed by mechanical elements in a tubular tube or barrel; the mixers contain a series of passive elements placed in a flow channel. These elements cause the plastic compound to subdivide and recombine in order to increase the homogeneity and temperature uniformity of the melt. There are no moving parts, and only a small increase in the energy is needed to overcome the resistance of the mechanical baffles. These mixers are located at the end of the screw plasticator, such as an injection molding machine, or before the screen changer and/or die of an extruder. In an extruder, if a gear pump is used, the static mixer is located between the screw and gear pump. They can be used to mix both different plastics and plastics with their component ingredients, such as color and additives.

Strain. The per unit change, due to force, in the size or shape of a body from its original size and shape. Strain is nondimensional but is usually expressed in units of length per units of length or percent. It is the natural logarithm of the ratio of gauge length at the moment of observation instead of the original cross-sectional area. Applicable to tension and compression tests.

Strength. The stress required to break, rupture, or cause a failure of a substance. Basically it is the property of a material that resists deformation induced by external forces. Maximum stress occurs when a material can resist the stress without failure for a given type of loading.

Stress. The intensity, at a point in a body (e.g., product, material, etc.), of the internal forces (or components of force) that act on a given plane through the point causing deformation of the body. It is the internal force per unit area that resists a change in size or shape of a body. Stress is expressed in force per unit area and reported in MPa, psi, and so on. As used in tension, compression, or shear, stress is normally calculated on the basis of the original dimensions of the appropriate cross section of the test specimen. This stress is sometimes called engineering stress; it is different than true stress, which takes into account the change in cross section.

Synergism. An arrangement or mixture of materials in which the total resulting performance is greater than the sum of the effects taken independently such as with alloying/blending plastics.

Syntactic. Identifies an orderly arrangement in a compound of components, ingredients, and so on, so that the product has absolutely isotropic mechanical properties.

Temperature controller, heating overshoot circuit. Used in temperature controllers to inhibit temperature overshooting on warm-up.

Temperature detector, resistance (RTD). An RTD contains a temperature sensor made from a material such as high-purity platinum wire. Resistance of the wire changes rapidly with the temperature. These sensors are about 60 times more sensitive than thermocouples.

Temperature measurement. Temperatures can be measured with a thermocouple or an RTD. Thermocouples tend to have shorter response time, while RTDs have less drift and are easier to calibrate. Traditionally, PID controls have been used for heating, and on-off controls have been used for cooling. From a temperature control point, the more recent use is the FLC. One of the FLC's major advantages is the lack of overshoot on startup, resulting in achieving the set point more rapidly. Another advantage is in its multivariable control where more than one measured input variable can effect the desired output result. This is an important and unique feature. With PID, one measured variable affects a single output variable. Two or more PIDs may be used in a cascade fashion, but with more variables they are not practical to use.

Temperature proportional-integral-derivative (PID). Pinpoint temperature accuracy is essential to be successful in many fabricating processes. In order to achieve it, microprocessor-based temperature controllers can use a PID algorithm acknowledged to be accurate. The unit will instantly identify varying thermal behavior and adjust its PID values accordingly.

Thermodynamic. It is the scientific principle that deals with the interconversion of heat and other forms of energy. Thermodynamics (*thermo* meaning *heat* and *dynamic* meaning *changes*) is the study of these energy transfers. The law of conservation of energy is called the first law of thermodynamics.

Thixotropic. A characteristic of material undergoing flow deformation where viscosity increases drastically when the force inducing the flow is removed. In respect to materials, it refers to those that are gel-like at rest but fluid or liquefied when agitated (such as during molding). It can also refer to having high static shear strength and low dynamic shear strength at the same time or losing viscosity under stress. It describes a filled plastic (e.g., bulk modling compounds [BMC]) that has little or no movement when applied to a vertical plane. Powdered silica and other fillers are used as thickening agents.

Tolerance, full indicator movement (FIM). FIM is a term used to identify tolerance with respect to concentricity. Terms used in the past were full indicator reading (FIR) and total indicator reading (TIR).

Torpedo. An unflighted cylindrical portion of the plasticator screw usually located at the discharge end that is providing additional shear heating capabilities for certain plastics.

Transistor. A semiconductor device for the amplification of currents required in different sensing instruments. The two principle types are field effect and junction.

Value analysis (VA). VA is the process of determining an amount regarded as a fair equivalent for something, that which is desirable or worthy of esteem, or a product of quality having intrinsic worth. Aside from technology developments, there is always a major emphasis on value-added services. It is where the fabricator continually tries to find ways to augment or reduce steps during manufacture with the target of reducing costs.

While there are many definitions of VA, the most basic is the following formula, where VA = (function of product)/(cost of the product). Immediately after the part goes into production, the next step that should be considered is to use the value engineering approach and the FALLO (follow all opportunities) approach. These approaches aim to produce products that meet the same performance requirements at a lower cost. If you do not take this approach, then your competitor will take the cost-reduction approach. VA is not exclusively a cost-cutting discipline. With VA, you literally can do "it all," which includes reducing costs, enhancing quality, and boosting productivity.

Vent purifier. The exhaust from vented plasticating barrels can show a dramatic cloud of swirling white gas; almost all of it is condensed steam proving that the vent is doing its job. However, a small portion of the vent exhaust can be other materials, such as byproducts released by certain plastics and/or additives, and could be of concern to plant personnel safety and/or plant equipment. Purifiers can be attached (with or without vacuum hoods located over the vent opening) to remove and collect the steam and other products. The purifiers include electronic precipitators.

Virgin plastic. Plastic materials in the form such as pellets, granules, powders, flakes, liquids, and so on that have not been subjected to any fabricating method or recycled.

Viscoelasticity. A material having this property is considered to combine the features of a perfectly elastic solid and a perfect fluid, representing the combination of elastic and viscous behavior of plastics.

Viscosity. The property of resistance to flow exhibited within the body of a material. In testing, it is the ratio of the shearing stress to the rate of shear of a fluid. Viscosity is usually taken to mean Newtonian viscosity, in which case the ratio of shearing stress to rate of shearing is constant. In non-Newtonian behavior, which is the usual case with plastics, the ratio varies with shearing rate. Such ratios are often called the apparent viscosities at the corresponding shear rates. Basically, this is the property of the resistance of flow exhibited within a body of material.

Viscosity, absolute. The ratio of shear stress to shear rate. It is the property of internal resistance of a fluid that opposes the relative motion of adjacent layers. Basically it is the tangential force on a unit area from either of two parallel planes at a unit distance apart, when the space between the planes is filled and one of the planes moves with unit velocity in its own plane relative to the other. The Bingham body is a substance that behaves somewhat like a Newtonian fluid in that there is a linear relation between rate of shear and shearing force, but it also has a yield value.

Viscosity, coefficient. It is the shearing stress necessary to induce a unit velocity gradient in a material. In actual measurement, the viscosity coefficient of a material is obtained from the ratio of shearing stress to shearing rate. This assumes the ratio to be constant and independent of the shearing stress, a condition satisfied only by Newtonian fluids. Consequently, in all other cases this includes plastics (non-Newtonian); values obtained are apparent and represent one point in the flow chart.

Viscosity, inherent. Refers to a dilute solution viscosity measurement where it is the ratio of the natural logarithm of the relative viscosity to the concentration of the plastic in grams per 100 ml of solvent.

Viscosity, intrinsic. Intrinsic viscosity is a measure of the capability of a plastic in solution to enhance the viscosity of the solution. Intrinsic viscosity increases with increasing plastic MW. It is the limiting value at infinite dilution of the ratio of the specific viscosity of the plastic solution to the plastic's concentration in moles per liter. Intrinsic-viscosity data is used in processing plastics. As an example, the higher intrinsic viscosity of injection-grade PET plastic can be extruded blow molded; this is similar to PETG plastic, which can be easily blow molded but is more expensive than injection-molded grade PET and PVC for blow molding.

Volumetric efficiency. The volume of plastic discharged from the machine during one revolution of the screw, expressed as a percentage of the developed volume of the last turn of the screw channel.

Vulcanization. Methods for producing a material with good elastomeric properties (rubber) involves the formation of chemical crosslinks between high-molecular-weight linear molecules. The starting polymer (such as raw rubber) must be of the noncrystallizing type, and its glass transition temperature T_g must be well below room temperature to ensure a rubbery behavior.

Weld line. Also called weld mark, flow line, or striae. It is a mark or line when two melt flow fronts meet during the filling of an injection mold.

X-axis. The axis in the plane of a material used as 0° reference. The y-axis is the axis on the plane of the material perpendicular to the x-axis, and the z-axis is the reference axis perpendicular to the x- and y-axes.

Y-axis. A line perpendicular to two opposite parallel faces.

Z-axis. The reference axis perpendicular to the x- and y-axes.

Further Reading

Abbott, W. H. *Statistics Can Be Fun*. Chesterland, OH: A. Abbott Publishing, 1980.
Acquarulo, L. A., and C. J. O'Neil. "Enhancing Medical Device Performance with Nanocomposite Polymers." *Medical Device and Diagnostic Industry*, May 2002.
Altshuler, T. L. *Fatigue: Life Predictions for Materials Selection and Guide*. Materials Park, OH: ASM Software, 1988.
Anderson, L. "SPI Adopts Trade Philosophy Statement." *Plastics Engineering*, March 2002.
Anderson, T. "Composite Fan Blades Safer than Metal." *Reinforced Plastics*, June 2002.
"Annual Market Data Book." *Plastics News*, December 1999–2010.
Arnold, K. G., et al. *Chemical Reviews*, vol. 57. 1959.
ASTM. *ASTM Book of Standards*, section 8: *Plastics*. West Conshohocken, PA: ASTM.
———. *ASTM Dictionary of Engineering Science and Technology*, 9th ed. West Conshohocken, PA: ASTM, 2000.
———. *ASTM Index—Annual Book of ASTM Standards*. West Conshohocken, PA: ASTM.
———. *ASTM International Directory of Testing Laboratories*. West Conshohocken, PA: ASTM, 1999.
"Auto Composites Set for Massive Growth." *Modern Plastics*, May 8, 2000.
"Auto Manifold." *Reinforced Plastics*, June 2001.
Avallone E., and T. Baumeister, eds. *Mark's Standard Handbook of Mechanical Engineers*. 10th ed. New York: McGraw Hill, 1996.
Baranek, S. L. "Software Package Increases Solid Modeling Capabilities." *MoldMaking Technology*, August 2002.
Bayer/Mobay. *Design Manual*. Bayer/Mobay: 1990.
Benedikt, G. M., and B. L. Goodall. *Metallocene-Catalyzed Polymers-Materials, Properties, Processing, and Markets*. Norwich, NY: Plastics Design Library, 1998.

Bernhardt, A., G Bertacchi, and A. Vignale. "Rationalization of Molding Machine Intelligent Setting and Control." *SPE-IMD Newsletter* 54, Summer 2000.

"Boeing Chief Predicts Faster, Smoother Flights in Next-Generation Airplane." *Design News*, June 2002.

Boonton Molding Co. "A Ready Reference for Plastics: Custom Molding Since 1921." *Boonton Molding Co. Bulletin*, 1933 through reprints to 1966.

Bozzelli, J. W. "Going from Hydraulic to Electric: Processor's Perspective." *Injection Molding*, 2002 annual.

———. *What Is Scientific Injection Molding?* Midland, MI: Injection Molding Solutions, 2002.

BP Publications. "Composite Material Throws Competition a Curved Ball." *Horizon* (BP publication), August 2002.

Brandi, D. L. "ISA SP95: The Factory to Business Link." *Industrial Compounding*, June 2000.

Bregar, B. "Magnesium Molding Gains Ground Globally." *Plastics News*, June 24, 2002.

Brinker, J. "Polymers Report Stress with Color Changes." *Design News*, July 2, 2001.

Bryce, D. "Why Offer Aluminum Molds for Production?" *Mold Making*, April 2002.

Burns and McDonnell Corp. "Ancient Engineering: Ideas and Methods That Have Endured 2000 Years of Progress." *Bench Mark Bulletin* 1, 1992.

Cappelletti, M. "In Defense of Plastics." *World Plastics Technology*, 2000.

Caraballo, W. "Flexibility and Customization Expand Packaging Options (Thermoform-Fill-Seal)." *Medical Device and Diagnostic Industry*, January 2002.

Cha, S., et al. "3-D Simulation of Thin-Wall Injection Molded Parts by CAE." *SPE-IMD Newsletter*, May 2002.

Chalmers, R. E. "Mold Steels." *Injection Molding*, June 2000.

Chamis, C. C., et al. *Hybrid Composites*. Reston, VA: American Institute of Aeronautics and Astronautics, 1979.

Cheremisinoff, N. P. *Polymer Characterization Laboratory Techniques and Analysis*. Norwich, NY: Plastics Design Library, 1996.

Ciullo, P. A. *The Rubber Formulary*. Norwich, NY: Plastics Design Library, 1999.

Colby, P. N. "Plasticating Components Technology: Screw and Barrel Technology." *Spirex Bulletin* (Spirex Corporation), 2000.

Coleman, B. D. "Thermodynamics of Materials with Memory Treatise." *Arch. Rat. Mech. Anal.*, 1964.

Colvin, R. "Automated Roll Handling Boosts Cast Film Operations; Cost Plays a Secondary Role to Improved Productivity, Quality, and Safety." *Modern Plastics*, November 2000.

"Computers: More Than Child's Play." *The Inside Line*, March 1999.

"Consumers' Computers Intolerable." *The Inside Line*, March 1999.

Corbmann, G. W. "Quality Audit vs. Certification Mania." *Kunststoffe*, March 2000.

Corripio, A. B. *Design and Application of Process Control Systems*. Research Triangle Park, NC: ISA, 1998.

Covas, J. A., J. F. Agassant, A. C. Diogo, J. Vlachopoulos, and K. Walters, eds. *Rheological Fundamentals of Polymer Processing*. Dordrecht, Netherlands: Kluwer, 1995.

Coxe, M., and C. M. F. Barry. "The Establishment of a Processing Window for Thin-Wall Injection Molding of Syndiotactic Polystyrene." Presented at the Society of Plastics Engineers Technical Conference (ANTEC), 2000.

Cramez, M. C., M. J. Oliveira, and R. J. Crawford. "Relationship between the Microstructure and the Properties of Rotational Molded Plastics." In *Coloring Technology for Plastics*, edited by R. M. Harris. Norwich, NY: Plastics Design Library, 1999.

Crawford, R. J., and J. L. Throne. *Rotational Molding Technology*, Norwich, NY: Plastics Design Library, 2002.

Cushion, R. F. *Engineers Malpractice*. Hoboken, NJ: Wiley, 1987.

Daido Steel. "Tool Steel Selection Software Targets Mold Making Startups." *Modern Mold*, April 1999.

Dealy, B. "Size Does Matter for Injection Units." *Modern Plastics*, July 2002.

Defosse, M. "Large Blow Molded Containers Take Off." *Modern Plastics*, April 2000.

———. "Material and Process Developments (Pultrusions) Promise New Applications." *Modern Plastics*, June 2002.

———. "Public's Education is Key to Vinyl's Acceptance." *Modern Plastics*, June 2002.

Dell'Arciprete, J., R. Malloy, and S. P. McCarthy. "Cavity Pressure Transfer Extends Prototype Tool Life." *Modern Plastics*, January 2000.

Deming, E. W. *Out of the Crisis*. Cambridge, MA: MIT Center for Advancement Engineering Studies, 1986.

Demirci, H. H. "Process Window Identification for a Very Tight-Tolerance Injection Molded Part with Multiple Performance Criteria." Presented at the Society of Plastics Engineers Technical Conference (ANTEC), 2001.

Dickens, P. R. Hague, and T. Wohlers. "Methods of Rapid Tooling Worldwide." *Moldmaking Technology*, October 2000.

Dispenza, J. "Reaching Out through Change." *SPE-IMD Molding Views*, May 2002.

Dorgham, M., and D. V. Rosato. *Designing with Plastic Composites*. Geneva, Switzerland: Interscience Enterprises, 1986.

Dostal, C. A. *Engineered Materials Handbook*, vol. 1: *Composites*. Materials Park, OH: ASM International, 1987.

———. *Engineered Materials Handbook*, vol. 2: *Engineering Plastics*. Materials Park, OH: ASM International, 1988.

Dowhower, K., and B. Bernhart. "Seals Upgrade Connectors." *Design News*, December 17, 2001.

Downey, J. P., and Pojman, J. A., eds. *Polymer Research in Microgravity*. Oxford: Oxford University Press, 2001.

Dratschmidt, F., et al. "Threaded Joints in Glass Fiber Reinforced Polyamide." *Journal of Polymer Engineering Science* 37(4): 744–55.

Dussault, R. "World Wired Web." *Motion Control*, April/May 2000.

Earle, J. H. *Engineering Design Graphics*. New York: Addison-Wesley, 1990.

Ebnesajjad, S. *Chemical Resistance*, 2nd ed., vols. 1 and 2. Norwich, NY: Plastics Design Library, 1994.

———. *Fluoroplastics*, vol. 1: *Non-Melt Processible Fluroplastics*. Norwich, NY: Plastics Design Library, 2000.

"Electric Machines and New Processes Catch Fire." *Plastics Technology*, January 2002.

"Electronics Molding Is on the Rebound." *Plastics Technology*, August 2002.

"Enabling Technology Developed for Cell Transplantation." *Medical Device and Diagnostic Industry*, May 2000.

"Energy." *World Plastics Technology*, 2001

Engelke, C., et al. "Putting Human Factors Engineering into Practice." *Medical Device and Diagnostic Industry*, July 2002.

Evans, B., et al. "Effective Use of Industrial Design in Rapid Product Development." *Medical Device and Diagnostic Industry*, September 1999.

Ezrin, M. "Plastics Analysis: The Engineer's Resource for Troubleshooting Product and Process Problems and for Competitive Analysis." *Plastics Engineering*, February 2002.

———. *Plastics Failure Guide: Cause and Prevention*. Cincinnati, OH: Hanser, 1996.

Ezrin, M., et al. "Case Studies of Plastics Failure Related to Improper Formulations." Presented at the Society of Plastics Engineers Technical Conference (ANTEC), 1999.

Farrell, R. E. "Artificial Intelligence in Injection Molding." *SPE-IMD Newsletter* 35, 1994.

Fasce, L. A., et al. *Fracture Behavior of Polypropylene Modified with Metallocene Catalyzed Polyolefin*. Norwich, NY: Plastics Design Library, 2001.

"FDA Clears Shell's PEN for Food Uses." *Plastics News*, May 8, 2000.

Fenichell, S. *Plastic: The Making of a Synthetic Century*. New York: Harper Business, 1996.

Finan, J. M. "Thermally Conductive Thermoplastics." *Plastics Engineering*, May 2000.

Firenze, A. R. *The Plastics Industry*. Hudson, MA: Adaptive Instruments, 2000.

Fitzgerald, M. "Using Simulation in Robotic Design." *Motion Control*, April/May 2000.

"Flexible Foaming Process for Refrigerators." *Plastics Technology*, January 2002.

"Flexible PVC." *Plastics Engineering*, June 2002.

Flick, E. W. *Epoxy Resins, Curing Agents, Compounds, and Modifiers: An Industrial Guide*, 2nd ed. Norwich, NY: Plastics Design Library, 1993.

———. *Plastics Additives*, 2nd ed. Norwich, NY: Plastics Design Library, 1993.

"Forget the Facts—Just Pass a Law." *Compressed Air*, July/August 1993.

Frankish, J., et al. "High-Performance Polymer/Metal Composite Replaces Lead." *Plastics Engineering*, October 2000.

Frantz, J. "How to Buy New RT Technology." *Moldmaking Technology*, August 2000.

Frederick, C. D., et al. *Rotational Molding*. Plastics Solutions International, 2000.

Freed, A. D. *Viscoplastic Model Development*. Washington, DC: NASA, 1995.

Freeman, R. "Pressure Forming with Style." *SPE Thermoforming Quarterly*, Spring 1999.

French, T. E., et al. *Engineering Drawing and Graphic Technology*. New York: McGraw-Hill, 1986.

"FR Nylons Get a Boost from Red Phosphorous." *Plastics Technology*, February 2002.

Fuges, C. "Rapid Prototyping: A Journey, Not a Destination." *MoldMaking Technology*, February 2000.

Garmabi, H., and M. R. Kamal. "Improved Barrier and Mechanical Properties of Laminar Polymer Blends." In *Imaging and Image Analysis Applications for Plastics*, edited by B. Pourdehimi. Norwich, NY: Plastics Design Library, 1999.

Gauthier, M. M. *Engineered Materials Handbook*, Desk Edition. Materials Park, OH: ASM International, 1995.

Genest, D. H. "Data Analysis Links Design with Manufacturability." *Moldmaking Technology*, August 2002.

Glanville, A. B. *Plastics Engineers Data Book*. New York: Industrial Press, 1971.

Goerz, R. "Improved Simulation of Barrier Screws." *Kunststoffe*, March 2000.

Goldman, A. Ya., et al. "Effect of Aging on Mineral-Filled Nanocomposites." In *Weathering of Plastics: Testing to Mirror Real Life Performance*, edited by G. Wypych. Norwich, NY: Plastics Design Library, 1999.

Goldsberry, C. "Good-Bye Wood, Hello Plastic." *Injection Molding*, June 2002.

———. "Purchasing Injection Molds: Buyers Guide." *Injection Molding*, 2000.

Gonzalez, M. "Special Rules Govern Design of Tools for Elastomers." *Mold Makers and Tooling*, March 1999.

Goodman, S. H. *Handbook of Thermoset Plastics*, 2nd ed. Norwich, NY: Plastics Design Library, 1998.

Gordon, D. "Aluminum Molds Go the Distance." *Molding Systems*, March 1999.

Grande, J. A. "Reformulated Acetals Meet Demands in Precision Gears, Auto Fuel Systems." *Modern Plastics*, May 2000.

Grimm, T. "Rapid Tooling Is Not the Future, It Is Today." *MoldMaking Technology*, February 2000.

Groleau, R. J. "Comparing Molding Machines Using Data Acquisition Equipment." Presented at the SPI Conference, June 7, 1994.

Handa, P., et al. "CO_2 Blown PETG Foams." In *Imaging and Image Analysis Applications for Plastics*, edited by B. Pourdehimi. Norwich, NY: Plastics Design Library, 1999.

Hannagan, T. "The Use and Misuse of Statistics." *Harvard Management Update*, May 2000.

Harley, S. "Road Safety: Driving Force to Reduce Deaths." *Horizon* (BP publication), 2002.

Harrington, J. P. *Who's Who in Plastics and Polymers*. Lancaster, PA: Technomic, 2000.

Harris, R. M. *Coloring Technology for Plastics*. Norwich, NY: Plastics Design Library, 1999.

Hauck, C., et al. "Auto Modules Will Play to Plastics' Strength." *Modern Plastics*, January 2000.

Haut, D. "Success by Design." *Medical Device and Diagnostic Industry*, September 1988.

"Hazardous Waste." *Compressed Air*, January 1989.

Heger, F. J. *Structural Plastics Design Manual*. Reston, VA: ASCE, 1981.

Heim, H. P., et al. *Specialized Molding Techniques*. Norwich, NY: Plastics Design Library, 2001.

Herrmann, J. "The Protection of Industrial Design." *Les Nouvelles* (Licensing Executives Society International publication), June 2000.

Hertzberg, R. W., et al. "Fatigue Testing: Flaws Make It Better." *Plastics World*, May 1977.

"High Hopes for Beer Bottles Enliven Packaging Conferences." *Plastics Technology*, January 2000.

Hoechst Celanese Corp. *Designing with Plastics: An Engineering Manual*. Dallas, TX: Hoechst Celanese, 1989.

Hull, J. *Compression and Transfer Molding*. New Britain, PA: Hull, 2000.

Hummel, M., et al. "Heater Selection." *Moldmaking Technology*, August 2000.

Hunkar, D. B. "Here's a Universal Scale for Grading Your Machine Capability." *Plastics Technology*, April 1993.

———. "MAD Approach to the Determination of Optimum Machine Settings in Injection Molding of Thermoplastics." Presented at the Society of Plastics Engineers Technical Conference (ANTEC), 1994.

Hunold, D., et al. "Injection Molding and Crosslinking." *Kunststoffe*, March 2000.

"Industry News: Dow Sharpens Cutting Edge." *Plastics Engineering*, February 2002.

"Input Device Developed for Medical Applications." *Medical Device and Diagnostic Industry*, February 2000.

International Organization for Standardization. *ISO Standards Compendium ISO 9000: Quality Management*, 9th ed. Geneva, Switzerland: ISO, 2001.

International Organization for Standardization. *ISO Standards Handbook: Statistical Methods for Quality Control*, 4th ed. Geneva, Switzerland: ISO, 1995.

"Intolerable Personnel Computer." *The Inside Line*, March 1999.

Jacob, A. "Spray-Up Offers Process Improvements." *Reinforced Plastics*, January 2002.

"Japan Takes Big Lead in Biocides Use in Plastics." *Modern Plastics*, May 2000.

Johnsen, T. "Test System Engineering for Medical Devices: A Guide." *Medical Device and Diagnostic Industry*, January 2002.

Johnson, B. P. "The Use of a Design of Experiments (DOE) to Optimize Processing Conditions of an Injection Molded Gear." Presented at the Society of Plastics Engineers Technical Conference (ANTEC), 2002.

Johnson, J. "Streamlining Polymer Selection for E/E Applications." *Plastics Engineering*, January 2002.

———. "Streamlining Polymer Selection for E/E Applications." *Plastics Engineering*, January 2002.

Johnson, L., et al. "Breathable TPE Films for Medical Applications." *Medical Device and Diagnostic Industry*, July 2000.

Joseph, D. *Internal Cooling Systems with New Capabilities*. Plastics Solutions International, 2000.

Juntgen, T., et al. "The Water Injection Molding Technique (WIT) as an Attractive Alternative and Supplement to Gas-Assisted Injection Molding (GAIM)." Presented at the Society of Plastics Engineers Technical Conference (ANTEC), 2002.

Kaszynski, J. "What Do You Know about Steel Quality?" *Moldmaking Technology*, August 2000.

Katz, H. "Forge of the Gods: An Eruption of New Biomaterials." *Medical Device and Diagnostic Industry*, July 2000.

Kaufman, M. *The First Century of Plastics*. London: Plastics Institute, 1963.

Keefer, L. B. "Creative before Capital Growth Strategy." *Moldmaking Technology*, August 2000.

"Keeping Up with Materials." *Plastics Technology,* January 2002.

Kelleher, B. "The Seven Deadly Sins of Medical Device Development." *Medical Device and Diagnostic Industry*, September 2001.

Kemmann, O., et al. "Simulation of the Micro Injection Molding Process." Presented at the Society of Plastics Engineers Technical Conference (ANTEC), 2000.

King, R. D., et al. "The Future Belongs to Biopolymers." *Modern Plastics*, January 2000.

King, R. W., et al. *The Effect of Nuclear Radiation on Elastomeric and Plastic Components: REIC Report No. 21*. Washington, DC: Defense Logistics Agency, 1961.

Klaus, B., et al. "A Technology in Transition with Electric Molding Technology." *SPE-IMD Newsletter* 52, Fall 1999.

Klempner, D., et al. *Advances in Urethane Science and Technology*. Shawbury, UK: Rapra, 2002.

Koenig, J. K. *Infrared and Raman Spectroscopy of Polymers*. Shawbury, UK: Rapra, 2002.

Krottner, V. "Teach Yourself Polishing." *Moldmaking Technology*, August 2000.

———. "Reclaiming the Lost Art of Benching." *Moldmaking Technology*, October 2000.

Kutz, M. *Mechanical Engineers' Handbook*. Hoboken, NJ: Wiley, 1998.

Landrock, A. H. *Handbook of Plastic Foams*. Norwich, NY: Plastics Design Library, 1995.

Langlois, M. "Super Injection Molding Makes Tiny Parts." *Job Shop Technology*, May 2000.

Larsen, A. N., et al. "Is the Shear Heating Phenomenon Truly Responsible for Viscosity Reduction in Thermoplastic Injection Molding?" Presented at the Society of Plastics Engineers Technical Conference (ANTEC), 1998.

Lauzon, M. "Wood-Fiber Uses Moving Beyond Decking." *Plastics News*, June 2002.

Leaversuch, R. "Fuel Cells Jolt Plastics Innovation." *Plastics Technology*, November 2001.

Leventon, W. "Medical Tubing Offers More (and Less) to Device Makers." *Medical Device and Diagnostic Industry*, January 2002.

———. "Part Making on a Very Small Scale: Micromolding for Medical Devices." *Medical Device and Diagnostic Industry*, May 2002.

Lubofsky, E. "Eye on Machine Vision." *IC Magazine*, January 2002.

Macosko, C. W. *Rheology: Principles, Measurements, and Applications*. Hoboken, NJ: Wiley, 1994.

Magonov, S. et al. "Atomic Force Microscopy, Part 6: Recent Developments in AFM of Polymers." *American Laboratory*, May 1998.

Maier, C., et al. *Polypropylene: The Definitive User's Guide and Databook*. Norwich, NY: Plastics Design Library, 1998.

Malloy, R. A. *Plastic Part Design for Injection Molding*. Cincinnati, OH: Hanser, 1994.

Mamzic, C. L. *Statistical Process Control*. Research Triangle Park, NC: ISA (International Standards Association), 1995.

Mapleston, P. "Blu-Ray Discs May Turn Acrylic's Prospects Around." *Modern Plastics*, August 2002.

———. "Cavity Pressure Control System." *Modern Plastics*, June 2002.

———. "End-Users Fire Up Demand for High Heat Thermoplastics." *Modern Plastics*, June 2002.

———. "Plastics Are Primed for Big Push in Auto Exteriors." *Modern Plastics*, July 2002.

Marks, W. F. "Stress-Strain Polynomials; Their Use Could Result in More Accurate Design Calculations." *DuPont Engineering Magazine*, Winter 1986.

Marmathy, S. "Design of Buildings for Fire Safety." *ASTM STP* 685, 1969.

Marsh, G. "Filling the Front-Line Training Gap." *Reinforced Plastics*, April 2002.

———. "Flying High with the Self-Build Movement." *Reinforced Plastics*, January 2002.
———. "MACT: A More Restrictive World for Boatbuilders." *Reinforced Plastics*, January 2002.
"Materials Data." *Plastics Engineering*, April 2002.
"Materials Update." *Injection Molding*, January 2002.
Maxwell, J. C. *The Philosophical Transactions of the Royal Society*, vol. 157. London: 1867.
Mazumdar, S. K. *Composites Manufacturing*. New York: CRC Press, 2002.
McConnell, V. P. "Composites and the Fuel Cell Revolution." *Reinforced Plastics*, January 2002.
Menges, G., M. Wenig, T. Folster. "Deformation Behavior of Thermoplastics for Non-Uniform Stress Distributions." *Kunststoffe*, 1990.
Menges, G., W. Michaeli, and P. Mohren. *How to Make Injection Molds*, 3rd ed. Cincinnati, OH: Hanser, 2001.
Michaeli, W., T. Folster, B. Lewen. "Beschreibung des Nichtlinear-Viskoelastischen Verhaltens mit dem Deforrnationsmodell." *Kunststoffe* 79, 1998.
Michaeli, W, U. Mohr-Matuschek, B. Lewen, T. Folster. "Kunststoffgereebtes Konstniieren." *Kunststoffe* 80, 1990.
"Microcellular Foam Technology." *Injection Molding*, December 2001.
"Micromolding for Microelectronics and Micro-Optics." *Injection Molding*, July 2000.
Miel, R. "Biocomposites Interest Grows." *Plastics News*, May 2000.
———. "Composite Boxes Ready to Carry Load." *Plastics News*, February 2000.
———. "Composite-Bodied Busses Hit the Road." *Plastics News*, February 2000.
———. "Plastics Use May Rise if Auto Voltage Changes." *Plastics News*, June 2002.
"Missing Link in Computer Integrated Manufacturing." *Injection Molding*, May 2000.
Moalli, J. *Plastics Failure: Analysis and Prevention*. Norwich, NY: Plastics Design Library, 2001.
Molinaro, H. "Colors Are Brewing for PET Beer Bottles." *Plastics Engineering*, February 2002.
Moore, S. "Innovative Reactive Extrusion Technologies Bring Added Value to Recycled Plastics." *Modern Plastics*, January 2000.
Mordfin, L. *Handbook of Reference Data for Nondestructive Testing*. West Conshohocken, PA: ASTM, 2002.
Mort, M. "Self-Assembling Circuitry-DNA." *Compressed Air*, September 1999.
Moss, M. "How to Work Better, Faster, and Cheaper." *Moldmaking Technology*, April 2002.
"Mulally Unveils Vision for Sonic Cruiser." *Design News*, September 17, 2001.
Murphy, J. "Flame Retardants." *Reinforced Plastics*, October 2001.
Murray, C. J. "Design with Lives in Mind." *Design News*, February 1990.
———. "Foam Exhibits Negative Poisson's Ratio." *Design News*, December 1989.
Murrill, P. W. *Fundamentals of Process Control Theory*, 3rd ed. Research Triangle Park, NC: ISA, 1999.
Nachtwey, P. "Motion Fundamentals: Fluid Power Basics." *Motion Control*, April/May 2001.
Nakason, C., et al. "Visualization of Polymer Melt Converging Flows in Extrusion." In *Imaging and Image Analysis Applications for Plastics*, edited by B. Pourdehimi. Norwich, NY: Plastics Design Library, 1999.
Nangrani, K., et al. "Effect of Pigments on the Flammability of Reinforced Thermoplastics." Presented at the Society of Plastics Engineers Technical Conference (ANTEC), 1987.

Neilley, R. "The Art of Fine-Tuning a Micro-Molding Business." *Injection Molding*, January 2002.

Newberry, A. L. "FEA Modeling Improves Its Worth in Composites Design." *Reinforced Plastics*, October 2000.

"New Materials and Processing = Design Success." *Design News*, April 23, 2000.

"New Specialty Polymers Improve Fuel Cell Economics." *Plastics Technology*, January 2002.

Ng, W. "Telecommunications (Fiber Optics) Conduit Is Growth Market for Pipe Processors." *Modern Plastics*, May 2000.

Ng, W., et al. "Suppliers Unveil New Approaches to Produce Controlled Rheology PP." *Modern Plastics*, April 2000.

"North American Plastics Recyclers and Brokers 2002 Survey." *Plastics News*, May 20, 2002.

Oberg, E., et al. *Machinery's Handbook*, 25th ed. New York: Industrial Press, 1996.

Ogando, J. "A Strong Door for Tough Times." *Design News*, November 5, 2001.

———. "Bonding Plastics 101." *Design News*, January 22, 2001.

———. "Guitar Assembly with No Strings Attached." *Design News*, August 6, 2001.

———. "Plastics Will Survive High-Volt Jolt." *Design News*, October 1, 2001.

Osswald, T., et al. *Injection Molding Handbook*. Cincinnati, OH: Hanser, 2002.

Owens-Corning Corp. *Fiberglas Plus Design: A Comparison of Materials and Processes*. Toledo, OH: Owens-Corning Corp., July 1985.

Ozburn, M. "States to Swap Glass for a PINTA." *Horizon* (BP publication), April 2002.

Partom, Y., and I. Schanin. "Modeling Nonlinear Viscoelastic Response." *Journal of Polymer Engineering Science* 22, October 1983.

"Plastics Additives and Compounding." *Elsevier International Magazine*, 2007.

"Plastics Additives in Europe." *Business Communications*, 2001.

"Plastics Applications." *Plastics Engineering*, April 2002.

Plastics Design Library. *The Effect of Creep-600 Graphs*. Norwich, NY: Plastics Design Library, 1991.

———. *The Effect of Sterilization Methods*. Norwich, NY: Plastics Design Library, 1994.

———. *The Effect of UV Light and Weather On Plastics and Elastomers*. Norwich, NY: Plastics Design Library, 1994.

———. *Fatigue and Tribological Properties of Plastics and Elastomers*. Norwich, NY: Plastics Design Library, 1995.

"Plastic Stints for Heart Patents." *Medical Device and Diagnostic Industry*, June 2002.

"PN Disposables Medical Supplies." *Plastics News*, May 2000.

Port, O. "In Transportation, One Word Plastics." *Business News*, March 6, 2000.

Portnoy, R. C. *Medical Plastics*. Norwich, NY: Plastics Design Library, 1998.

Pourdehimi, B. *Imaging and Image Analysis Applications for Plastics*. Norwich, NY: Plastics Design Library, 1999.

"PP Is on the Fast Track in Automotive." *Plastics Technology*, January 2002.

Price, P. "Nanotechnology: The Next Big Thing is Really Small." *Horizon* (BP publication), April 2002.

Pryweller, J. "Redesigning the Wheel." *Plastics News*, March 2002.

"Rapid Prototyping, Rapid Tooling." *Plastics Engineering*, April 2002.

"Recycling at K 2001." *Plastics Technology*, February 2002.

Reinfrank, G. B., et al. "Molded Glass Fiber Sandwich Fuselage (etc.) for BT-15 Airplane." Army Air Force Tech. Report No. 5159, November 8, 1944.

Renstrom, R. "Bistro Boasts Horse of a Different Material." *Plastics News*, March 18, 2002.

———. "JSF Flying on More Composites." *Plastics News*, July 1, 2002.

———. "Sliding Zippers Find Favor Among Packagers." *Plastics News*, March 18, 2002.

"Resin in Electronics." *Plastics News*, July 22, 2002.

Rieger, N. F., et al. "The Basics of Finite Element Modeling." *Machine Design*, April 9, 1981.

Roberts, G. *Tool Steels*, 5th ed. Materials Park, OH: ASM, 1998.

Rohsenow, W. M., et al. *Handbook of Heat Transfer*, 3rd ed. New York: McGraw-Hill, 1998.

Rosato, D. V. "Advanced Engineering Design Short Course." Presented at the American Society of Mechanical Engineers Conference, 1983.

———. *Asbestos: Its Industrial Applications*. Dordrecht, Netherlands: Kluwer, 1959.

———. *Blow Molding Handbook*. Cincinnati, OH: Hanser, 1989.

———. *Blow Molding Handbook*, 2nd ed. Cincinnati, OH: Hanser, 2002.

———. *Concise Encyclopedia of Plastics*. Dordrecht, Netherlands: Kluwer, 2000.

———. *Current and Future Trends in the Use of Plastics for Blow Molding*. New York: American Society of Mechanical Engineers, 1990.

———. "Design Features That Influence Part Performance." *SPE-IMD Newsletter* 46, 1997.

———. *Designing with Plastics*. Dordrecht, Netherlands: Kluwer, 2001.

———. *Designing with Plastics and Composites*. Dordrecht, Netherlands: Kluwer, 1991.

———. "Designing with Plastics." *Medical Device and Diagnostic Industry*, July 1983.

———. *Designing with Reinforced Composites*. Cincinnati, OH: Hanser, 1997.

———. *Extruding Plastics: Practical Processing Handbook*. Dordrecht, Netherlands: Kluwer, 1998.

———. *Filament Winding*. Hoboken, NJ: Wiley, 1964.

———. *From Laminates to Composites*. New York: Cahners, 1968.

———. "Industrial Plastics in Materials Handling." *International Mgm. Soc.*, October 1985.

———. "Injection Molding." In *Guide to Short Fiber Reinforced Plastics*, edited by R. F. Jones. Cincinnati, OH: Hanser, 1998.

———. *Injection Molding Handbook*, 3rd ed. Dordrecht, Netherlands: Kluwer, 2000.

———. "Injection Molding Higher Performance Reinforced Plastic Composites." *Journal of Vinyl and Additive Technology*, September 1996.

———. "Injection Molding in the 21st Century." *SPE-IMD Newsletter* 53–54, 2000.

———. *Markets for Plastics*. Dordrecht, Netherlands: Kluwer, 1969.

———. "Materials Selection." In *Encyclopedia of Polymer Science and Engineering*, vol. 9, edited by H. Mark, N. Bikales, C. G. Overberger, G. Menges, and J. I. Kroschwitz, 337–79. Hoboken, NJ: Wiley, 1987.

———. "Materials Selection and Reinforced Plastics: Thermosets." In *Concise Encyclopedia of Polymer Science and Engineering*, edited by J. I. Kroschwitz. Hoboken, NJ: Wiley, 1990.

———. "Materials Selection: Polymeric Matrix Composites." In *The International Encyclopedia of Composites*, edited by S. M. Lee. New York: VCH, 1991.

———. "Non-Woven Fibers in Reinforced Plastics." *Industrial and Engineering Chemistry* 54(8): 30–37.

———. "Nose Cone of First US Moon Vanguard Rocket is Made in Manheim" *New Era* (Lancaster, PA newspaper), November 30, 1957.

———. "Notes." *SPE-IMD Molding Views*, May 2002.

———. "Outer Space Parabolic Reflector Energy Converters." *Society for the Advancement of Material and Process Engineering*, June 1963.

———. *Plastic Engineered Product Design*. Amsterdam: Elsevier, 2003.

———. *Plastic Product Material and Process Selection Handbook*. Amsterdam: Elsevier, 2004.

———. "Plastic Replaces Aorta, Permits Living Normal-Long Life." Newton, Massachusetts: Newton-Wellesley Hospital, March 1987.

———. "Plastics and Solid Waste." Providence: Rhode Island School of Design, October 1989.

———. *Plastics Industry Safety Handbook*. New York: Cahners, 1973.

———. "Plastics in Missiles." *British Plastics*, August 1960.

———. "Plastics in the Wire and Cable." *PW*, June 1967.

———. *Plastics Processing Data Handbook*, 2nd ed. Dordrecht, Netherlands: Kluwer, 1997.

———. "Polymers, Processes, and Properties of Medical Plastics." In *Synthetic Biomedical Polymers*, edited by M. Szycher and W. J. Robinson. New York: CRC Press, 1980.

———. "Radomes." *Electronic Design News*, 1963

———. *Reinforced Plastics Handbook*. Amsterdam: Elsevier, 2005.

———. *Rosato's Plastics Encyclopedia and Dictionary*. Cincinnati, OH: Hanser, 1993.

———. "Target for Zero Defects." *SPE-IMD Newsletter* 26, 1991.

———. "Theoretical Potential for Polyethylene." USAF Materials Lab., WPAFB, 1944.

———. "Thermoset Polymers." In *The Encyclopedia of Packaging Technology*, 2nd ed., edited by A. L. Brody and K. S. Marsh. Hoboken, NJ: Wiley, 1997.

———. "Thermosets." In *The Encyclopedia of Polymer Science and Engineering*, vol. 14, edited by H. Mark, N. Bikales, C. G. Overberger, G. Menges, and J. I. Kroschwitz, 350–91. Hoboken, NJ: Wiley, 1988.

———. "Weighing Out the Aircraft Market—It's New Plastics that Count." Paper 68-320, American Institute of Aeronautics and Astronautics, Palm Springs, CA, April 1968.

———. "What Molders Must Do about ANSI Safety Specifications." *PW*, April 1978.

Rosato, D. V., et al. "Non-Metallic Composite Materials and Fabrication Techniques Applicable to Present and Future Solid Rocket Bodies." Presented at the ARS Conference, Salt Lake City, February 1961.

Rosato, D. V., and G. Lubin. "Application of Reinforced Plastics." Presented at the Fourth International Reinforced Plastics Conference, British Plastics Federation, London, November 25–27, 1964.

Rucinski, P. "Manufacturing Solutions for the Automation, Control, and Monitoring of the Injection Molding Process." *SPE-IMD Newsletter* 59, Spring 2002.

Rudd, L. E. "Blood Flows through Artificial Vessels." *Design News*, August 6, 2001.

Rupprecht, L. *Conductive Polymers and Plastics in Industrial Applications*. Norwich, NY: Plastics Design Library, 1999.

Saab, M. A. "Applications of High Pressure Balloons in the Medical Device Industry." *Medical Device and Diagnostic Industry*, September 2000.

Saile, R. "Understanding Thermoplastic Part Warpage." *Molding Systems*, June 2000.

Salerni, C. M. "Light-Cured Cyanoacrylates: An Adhesive Option for Medical Device Assembly." *Medical Device and Diagnostic Industry*, June 2002.

———. "Selecting Engineering Adhesives for Medical Device Assembly." *Medical Device and Diagnostic Industry*, June 2000.

"Santoprene TPE Absorbs the Shock in Ergonomic Hammers." *Modern Plastics*, April 2000.

Sanze, J. "Biaxial Foamed Films Offer Broad Properties Palette." *Modern Plastics*, August 2002.

Scaeberle, M., et al. "Raman Chemical Imaging: Noninvasive Visualization of Polymer Blend Architecture." *Analytical Chemistry* 67, 1995.

Schindler, B. M. "Made in Japan, W. Edwards Deming." *ASTM Standardization News*, February 1983.

Schmachtenberg, E. "Mechanical Properties of Nonlinear Viscoelastic Materials." PhD diss., RWTH, Aachen, Germany, 1985.

Schott, N., et al. "Optimization in Process Control for Uniform Quality of the Optical Components." Presented at the Society of Plastics Engineers Technical Conference (ANTEC), 2001.

Schut, J. H. "Long-Glass Leader: Thermoplastic Composites." *Plastics Technology*, August 2002.

———. "Big German Plant May Relieve U.S. Bottleneck in Recycling Carpet Nylon." *Plastics Technology*, May 2002.

———. "Foamed Films Find New Niches." *Plastics Technology*, February 2002.

Schwartz, R. T., and D. V. Rosato. "Structural Sandwich Construction." In *Composite Engineering Materials*, edited by A. G. H. Dietz, 165–81. Cambridge, Massachusetts: MIT Press, 1969.

Schwartz, R. T., and H. S. Schwartz. *Fundamental Aspects of Fiber Reinforced Plastic Composites*. Hoboken, NJ: Wiley, 1968.

Sepe, M. *Dynamic Mechanical Analysis*. Norwich, NY: Plastics Design Library, 1998.

Shay, R. M. "Estimating Linear Shrinkage of Semicrystalline Resins from Pressure-Volume Temperature (pvT) Data." *SPE-IMD Newsletter* 49, Fall 1998.

Shepard, S. "Eli Whitney: The Controversial Inventor." *Product Engineering*, October 25, 1965.

Sherman, L. M. "Metallocene VLDPE is a Tough New Contender for Flexible Packaging." *Plastics Technology*, January 2002.

———. "Polyurethane: Get Ready for HCFC Phase-Out." *Plastics Technology*, December 2001.

———. "The New Polypropylenes." *Plastics Technology*, May 2002.

Sims, F. *Engineering Formulas*. New York: Industrial Press, 1999.

Sims, G. *Composite Testing*. Plastics Solutions International, 2000.

Sippel, M. "Sensors Emerge as Machine Vision Tools." *Design News*, September 17, 2001.

Snyder, M. R. "Milacron Launches Internet-Based Link for Injection Molding Machine Diagnostics." *Modern Plastics*, July 2000.

———. "Robots Fill Greater Role in Integrated Automation Systems." *Modern Plastics*, April 2000.

Society of Plastics Engineers. *Plastics Engineering Guide on Extrusion Technology and Troubleshooting*. Newtown, CT: Society of Plastics Engineers, 2001.

Society of the Plastics Industry. *Facts and Figures of the U.S. Plastics Industry: Annual updates*. Washington, DC: Society of the Plastics Industry, 2010.

———. "Plastics Industry Report." *Plastics Engineering*, April 2002.

"SPC System Meets Custom Applications." *Injection Molding*, July 2002.

Spence, T., et al. "Rheology Studies." *SPE Thermoforming Quarterly*, Spring 1999.

"SPS Takes Early Application in Medical." *Modern Plastics*, July 2000.

SRI Consulting. "Biodegradable Polymers." *Chemical Industry Newsletter* 1, 2000.

"Stack Mold Now Performs In-Mold Assembly." *Plastics Technology*, August 2002.

Sterbenz, C. E. "One Word: Plastics." *Providence Journal*, July 2, 2000.

Szycher, M. *Biocompatible Polymers, Metals, and Composites*. Lancaster, PA: Technomic, 1983.

Szycher, M., and W. J. Robinson. *Synthetic Biomedical Polymers*. Lancaster, PA: Technomic, 1980.

Tai, J. "Coarse-Fine Method for Tension Control." *Motion Control*, October/November, 2000.

Tcharkhtchi, A. "Examining Elliptical Surface Defects on Angioplasty Balloons." *Medical Device and Diagnostic Industry*, May 2002.

"Testing Against Testing Against Trouble." *World Plastics Technology*, 2001

"There Oughta Be a Law: Consumer Abuse." *Consumer Reports*, October 1999.

"Thermoformed (Coextruded Plug-Assisted Twin-Sheet) Fuel Tanks Show Promise." *Modern Plastics*, April 2000.

"Thermoplastics in Major Markets." *Plastics Engineering*, May 2000.

"Thermosets 2000." *Injection Molding*, June 2000.

"These Fuel Tanks Flex." *Design News*, March 26, 2002.

Thirlwell, J. "Seven Key Advantages of Hot Runner Systems." *MoldMaking Technology*, April 2002.

Thompson, S. *Handbook of Mold, Tool, and Die Repair Welding*. Norwich, NY: Plastics Design Library, 1999.

Throne, J. L. *Technology of Thermoforming*. Cincinnati, OH: Hanser, 1996.

"Toothpaste in Tubes 100 Years and Still Going: Tube Council Of North America." *Tube Topics* 26(3), 1992.

"Toyota Will Use SMC Boxes for Truck Beds." *Modern Plastics*, August 2002.

Trantina, G. G., et al. "Standardization: Is it Leading to More Relevant Data for Design Engineers?" Presented at the Society of Plastics Engineers Technical Conference (ANTEC), 1994.

"Trexel Licenses Users of Demag Ergocell System." *Plastics Technology*, December 2001.

Trzaskoma, P. P., et al. *Characteristics of Rigid Polymer Foams as Related to their Use for Corrosion Protection in Enclosed Metal Spaces*. Houston, TX: NACE, 1999.

Tustison, L. "Overmolding Overview." *Injection Molding*, July 2002.

"Two New Ideas for Injection Molding Unveiled at ANTEC Meeting." *Plastics Technology*, August 2002.

U.S. Superintendent of Documents. *Advanced Composites Design Guide, MIL-HDBK-5C*. Washington, DC: Government Printing Office, 1980.

———. *Plastics for Aerospace Vehicles, MIL-HDBK-17A and 17B*. Washington, DC: Government Printing Office, 1981.

———. *Structural Plastics Design Manual, ASCE No. 63 or No. 023-000-00495-0*. Washington, DC: Government Printing Office, 1982.

———. *Structural Sandwich Composites, MIL-HDBK-23A*. Washington, DC: Government Printing Office, 1974.

Vinyl Institute. *Environmental Briefs*. Alexandria, VA: Vinyl Institute, May 2000.

Watkins, F., et al. "Control Technologies for Safety and Productivity." *InTech ISA*, February 2002.

Wenskus, J. J. "Transfer Point Control Comparison between Mold Parting Line and the Standard Strategies." *SPE-IMD Newsletter 22*, Fall 1989.

Wetton, R. E., et al. "Comparison of Dynamic Mechanical Measurements in Bending, Tension, and Torsion." Proceedings of the Society of Plastics Engineers Technical Conference (ANTEC), May 1989, pp. 1160–62.

White, J., et al. *Polymer Mixing: Technology and Engineering*. Cincinnati, OH: Hanser, 2001.

Wigotsky, V. "Compounding." *Plastics Engineering*, January 2002.

Wigotsky, V. "Plastics Testing." *Plastics Engineering*, February 2002.

Wille, D. "Producing Bubble/Taper Tubing for Medical Applications." *Medical Device and Diagnostic Industry*, January 2000.

Wilson, D., et al. *Composite Design Manual*. Newark, DE: University of Delaware, 1980.

Winkler, K. "Web-Based Collaboration for Plastic Medical Devices." Presented at the Society of Plastics Engineers Technical Conference (ANTEC), May 8, 2002.

Woishnis, William. *Permeability and Other Film Properties of Plastics and Elastomers*. Norwich, NY: Plastics Design Library, 1995.

Wright, D. C. *Environmental Stress Cracking of Plastics*. Norwich, NY: Plastics Design Library, 1996.

Wulpi, D. J. *Understanding How Components Fail*, 2nd ed. Materials Park, OH: ASM, 1999.

Wypych, G. *Handbook of Fillers*, 2nd ed. Norwich, NY: Plastics Design Library, 1999.

———. *Handbook of Solvents*. Norwich, NY: Plastics Design Library, 2001.

———. *Weathering of Plastics: Testing to Mirror Real Life Performance*. Norwich, NY: Plastics Design Library, 1999.

Young, W. C. *Roark's Formulas for Stress and Strain*, 6th ed. New York: McGraw-Hill, 1998.

Zahnel, K., et al. *Introduction to Microcontrollers*. Oxford, UK: Butterworth-Heinemann, 1997.

Zhao, C. H., et al. "A Fuzzy Rule Based Automatic V/P Transfer System for Thermoplastic Injection Molding." Presented at the Society of Plastics Engineers Technical Conference (ANTEC), 1997.